T0074216

"*Environmentalism from Below* is a much-needed and important book. In it, Dawson goes beyond narrow and technocratic imaginaries rooted in the nation-state, but also takes us past abstract romantic appeals to clearly trace the emancipatory potentials of global peoples' environmental movements. Carefully researched and accessibly written, the book connects food, cities, energy, conservation, debt, and borders in a narrative that manages to be both a sharp wake-up call and an optimistic assessment of what our common liberation can look like. This book is a must-read for anyone who feels there must be more to environmental justice than climate accords."

—GIANPAOLO BAIOCCHI, Director of the Urban Democracy Lab,
New York University, and author of *We, the Sovereign*

"*Environmentalism from Below* offers a politically erudite and passionate cacophony of momentum drawing from the world's variegated yet articulated grassroots, all attempting in solidarity to upend the transgression of key planetary ecological relations. Deploying an intersectional form of analysis and mobilization, the book powerfully examines the interplay among how food is produced, cities inhabited, space enclosed, and energy generated in an effort to abolish the debilitating indebtedness of the majority to capital's voracious calculations and their entrapment amidst borders. The book embodies the exigencies for the synergies of multiple movements underway—of people, affordances, collective capacities, rights, and resources—toward more just dispositions and the prospect of attaining a livable world."

—ABDOUMALIQ SIMONE, Senior Professorial Fellow
at the Urban Institute, University of Sheffield, and author
of *The Surrounds: Urban Life Within and Beyond Capture*

"*Environmentalism from Below* brims with fresh insights and new approaches to some of the most vexing issues of our time. In lucid, passionate prose, Ashley Dawson charts the global alliances forged from below against unregulated plunder and ecocide. Few scholars can match Dawson's vast transnational experience as an environmental scholar-activist. His global yet textured understanding of resistance movements from Bolivia, South Africa, India, Brazil, the US, and far beyond makes this a profound contribution to our understanding of how common struggles are forged. *Environmentalism from Below* is sure to become a staple in the environmental classroom as well as a guiding light for activists."

—ROB NIXON, Barron Family Professor of Environment and Humanities,
Princeton, and author of *Slow Violence and the Environmentalism of the Poor*

"Ashley Dawson takes us on a wondrous tour of communities working for life after capitalism. These grassroots ecologies are so potent, their promise so profound, they've elicited lethal violence from the state and private sector. For that reason, *Environmentalism from Below* is also an atlas of the world's most important struggles."

—RAJ PATEL, Research Professor, University of Texas at Austin,
and coauthor of *Inflamed: Deep Medicine and the Anatomy of Injustice*

"On a global scale humankind faces multiple existential crises as a result of climate change and the systematic environmental degradation that has brought us collectively to the brink. Populations in the Global South are most at risk, owing to decades of austerity measures imposed on peasant and Indigenous communities by the cruel alliances of neocolonial and neoliberal authoritarian governments, transnational corporations, and a host of multilateral NGOs. Ashley Dawson reframes these grim realities to emphasize how grassroots communities proactively resist the privatization and toxic exploitation of the natural world in innovative and empowering ways. Altogether, their examples stand as road maps for what many more of us will likely face in coming years."

—DINA GILIO-WHITAKER, author of *As Long as Grass Grows: The Indigenous Fight for Environmental Justice from Colonization to Standing Rock*

"True to its aim, this book celebrates ideas and actions that come from below. It is a book that deserves to be celebrated as it presents clear evidence of active organizing and resistance by climate victims and the dispossessed against manifestations of neocolonial and oppressive policies and actions. *Environmentalism from Below* is a book that fossil fuel tycoons and other purveyors of fictional environmental optimism will hate."

—NNIMMO BASSEY, author of *To Cook a Continent: Destructive Extraction and the Climate Crisis in Africa*

"Though debates rage on the climate Left about what language or strategy ought to be taken in order to confront the climate crisis, those involved in such conversations frequently seem to have their minds in the clouds and no grounded connection to existing class struggles. Dawson stands these critics on their heads by foregrounding the wildly diverse, actually existing, and ineluctably global people's movements for climate justice. In these scattered movements of urban squatters, migrants, industrial workers, peasant farmers, feminists, and Indigenous nations, one finds more comprehensive strategies for confronting imperialism and capitalism, which are the roots of environmental crises. *Environmentalism from Below* is a readable, practical, and inspiring guide to building ecological counterpower."

—KAI BOSWORTH, author of *Pipeline Populism: Grassroots Environmentalism in the 21st Century*

"Ashley Dawson's book focuses on environmentalism from below and enlightens us on all those central issues, such as the food model, agroecology, the debates on the just energy transition, the question of the sustainability of life in big cities, and climate debt. Written with commitment and elegance, this is an indispensable book for understanding the re-existence process and the organizational fabric, especially in the Global South."

—MARISTELLA SVAMPA, Ecosocial and Intercultural Pact from the South

# ENVIRONMENTALISM FROM BELOW

## HOW GLOBAL PEOPLE'S MOVEMENTS ARE LEADING THE FIGHT FOR OUR PLANET

Ashley Dawson

Haymarket Books
Chicago, Illinois

*Author proceeds from sales of this book will go to some of the grassroots organizations profiled in these pages, including groups fighting against extractivism, for agroecology, and for people's right to the city.*

Published in 2024 by
Haymarket Books
P.O. Box 180165
Chicago, IL 60618
773-583-7884
www.haymarketbooks.org
info@haymarketbooks.org

ISBN: 979-8-88890-058-1

Distributed to the trade in the US through Consortium Book Sales and Distribution (www.cbsd.com) and internationally through Ingram Publisher Services International (www.ingramcontent.com).

This book was published with the generous support of Lannan Foundation, Wallace Action Fund, and Marguerite Casey Foundation.

Author proceeds from sales of this book will go to some of the grassroots organizations profiled in these pages, including groups fighting against extractivism, for agro-ecology, and for people's right to the city.

Special discounts are available for bulk purchases by organizations and institutions. Please email info@haymarketbooks.org for more information.

Cover artwork and design by Josh MacPhee.

Library of Congress Cataloging-in-Publication data is available.

Entered into digital printing December, 2023.

# CONTENTS

# A GLOBAL PEOPLE'S MOVEMENT

I arrived in the Bolivian city of Cochabamba in mid-April of 2010 with a delegation of environmental justice activists from New York City. We had traveled to Bolivia to participate in the World People's Conference on Climate Change and the Rights of Mother Earth, a global gathering of climate justice activists convened by Bolivian president Evo Morales in response to the United Nations climate summit in Copenhagen the previous December.[1] The meeting in Copenhagen, the fifteenth Conference of Parties to the United Nations Framework Convention on Climate Change (COP15), had been a dismal failure. Negotiators from wealthy nations met behind closed doors to jettison their commitments to substantial reductions in greenhouse gas emissions, cuts made mandatory in the Kyoto Protocol of 1997.

Supported by a bloc of other progressive Latin American governments, the Cochabamba meeting was held less than half a year after the debacle in Copenhagen. The location was a significant one: a decade earlier, popular social movements in Cochabamba had triumphed against neoliberal privatization schemes and repressive state power in what became known as the "Water Wars."[2] It was against this backdrop of popular anti-capitalist mobilization—juxtaposed with elite failure to address the climate catastrophe at COP15—that the World People's Conference on Climate Change and the Rights of Mother Earth took place.

Foremost on my mind as I headed to Cochabamba for the gathering was the question of what an environmental liberation movement that began from the needs of the peoples of the Global South would look like. It seemed clear that an ecological, economic, political, and cultural reconstruction of the Global North nations could not take place without a similar transformation of the poorer countries. And yet it was precisely the perspectives and needs of people in peripheral nations—as well as the majority of people in the core nations—that had been systematically silenced and excluded in official climate summits like the one in Copenhagen. How, I wanted to know, could the peoples of the Global South play a role as protagonists of the struggle for ecological reconstruction? What, at bottom, do the governments of the wealthy imperial countries owe the peoples of the Global South, and how might these debts best be paid?

The gathering of climate justice activists in Cochabamba generated what for me remain some of the most radical and important responses to the question of how to stop capitalism and colonialism from destroying the ecological conditions necessary to the survival of humanity—and most other species of life on the planet to boot. Drawing on the lessons I learned in Cochabamba, this book sets out to imagine what global ecological reconstruction might look like if it were carried out by and for those who are most vulnerable to—but also least responsible for—the climate crisis. These also happen to be the people who have not been wholly divorced by the capitalist system from sustainable ways of living and worldviews that make such balanced lives possible.

The industrialized nations in the Global North are responsible for 92 percent of excess carbon emissions.[3] These rich nations have effectively colonized the atmospheric commons. And it's not just a historical problem: every year, the average person in the US, Canada, or Australia emits roughly fifty times more carbon than the average person in a country like Mozambique.[4] Although US President Barack Obama had made the promise of a Green New Deal part of his first electoral campaign in 2008, the US delegation at COP15 lobbied to kill the Kyoto agreement, pushing instead for a loose framework in which individual nations set their own purely voluntary emissions targets—the prologue to the ineffectual Paris Agreement of 2015.[5]

The liquidation of the Kyoto Protocol's legally mandated emissions cuts and the principle of "common but differentiated responsibilities" that undergirded them was a severe blow to the aspirations of Global South nations and frontline communities to contain dangerous global warming. While men in suits negotiated the future of the planet, climate activists demanding "System Change, Not Climate Change!" outside in the streets of Copenhagen in 2009 were met with heavy-handed police repression. It was clear that the elite bureaucrats and corporate leaders meeting in swank conference halls would brook no public dissent.[6]

The alarming implications of the negotiations in Copenhagen were immediately clear to delegates from the Global South. Lumumba Di-Aping, Sudanese leader of the G-77 bloc of poor countries, argued that negotiators for the wealthy nations were asking their colleagues from the Global South to "sign a suicide pact."[7] Working outside the UN negotiating process, the Obama administration had contrived a non-accord and threatened to deny climate aid to developing countries that refused to sign on to the deal. In response, Di-Aping called the 2°C warming that the rich nations considered acceptable "certain death for Africa," and said that a type of "climate fascism" was being imposed on Africa by high emitters.

Echoing Di-Aping's words, Evo Morales, Bolivia's first Indigenous president, spoke out during a meeting with European social movements at the Copenhagen summit. For Morales, the refusal of official delegations from the world's core industrialized nations to agree to binding emissions cuts underlined that it was up to popular movements to mobilize for climate justice. Reporting back on the progress of official negotiations, Morales said:

> There are those who defend capitalism and therefore defend the culture of death. We defend socialism, and therefore, we are on the side of the Culture of Life. Since we are not going to have the power to define the destiny of the world and humanity at the level of the heads of state, then we propose that the organized people of the world decide the destiny of humanity and the future of the planet.[8]

Capitalism or life. Elite state and corporate power versus popular power. These stark oppositions are as true today as they were in 2009. These oppositions were posed not simply by leaders of Global South countries already coping over a decade ago with inundation from rising seas and the melting of life-giving sources of drinking water, but also by climate science itself. Despite abundant evidence that business as usual would lead to catastrophe, powerful nations and corporations have spent years proffering false, ineffectual solutions like carbon trading and offsetting, not to mention increased border controls and greater militarization, which has resulted in thousands of deaths of climate refugees fleeing from problems created by the rich. Indeed, from the first gathering in 1995 in Berlin, the COP meetings have been organized around market-based initiatives demanded by the US and other wealthy nations.

To take but one example, the so-called Clean Development Mechanism (CDM) was originally proposed by Brazil as a means through which industrialized countries would pay penalties for excessive pollution, fines that could be used to bankroll mitigation and adaptation projects in the Global South. But under pressure from rich nations, the CDM was transformed into a market-based program that allows polluting corporations and countries to implement "clean" projects such as large-scale plantations, wind farms, and mega-dams in poorer nations to "offset" their own emissions-reduction obligations.

These programs provide a fig leaf to multinational corporations, allowing them to continue making money while avoiding any significant change to their polluting practices. They have not cut carbon emissions significantly, but have dispossessed people in the poorer countries of their land and livelihoods.[9] Climate justice activists such as Tom Goldtooth of the Indigenous Environmental Network has argued at counter-summits to the UN Climate Meetings that these schemes are actually "scams," "financial laundering mechanisms" that cause conflict and promote a fresh round of enclosure, colonialism, and genocide among the peoples of the Global South.[10] For Goldtooth, the climate crisis is only the latest symptom of a colonialist and capitalist system defined by a catastrophic imperative to grow ceaselessly on a finite planet.

Throughout this book I learn from and help articulate what I call *environmentalism from below*. In framing this concept, I draw on the work of Ramachandra Guha and Joan Martinez-Alier, environmental historians and activists who explain that "the environmentalisms of the poor originate in social conflicts over access to and control over natural resources" that date back to the colonial era in places like South Asia and Latin

America.[11] Contrary to arguments that tend to see environmentalism as a perquisite of affluent societies no longer burdened with having to meet basic material needs, Guha and Martinez-Alier argue that "environmentalisms of the poor" manifest as efforts to "retain under their control the natural resources threatened by state takeover or by the advance of the generalized market system."[12] Guha and Martinez-Alier's work refers explicitly to the efforts of Indigenous people such as India's Adivasis to prevent the colonial and postcolonial state from appropriating their forests, acts of enclosure often carried out in the name of resource conservation. The scholars' focus on protests by people who depend on local resources for survival could be extended far beyond India, to struggles by environmental defenders who are being murdered in increasing numbers every year for trying to protect the natural resources they depend on.

Guha and Martinez-Alier framed their model of "environmentalism of the poor" in relation to relatively isolated protests in the past. Today, however, the dispossession of the environmental commons is a global phenomenon. The peril of ecological collapse is unfolding on a planetary scale, generating immense tragedies but also the potential for new solidarities and modes of resistance.[13] Environmentalism from below is animated by struggles for collective control of the environmental and social commons in the face of global environmental degradation and dispossession carried out by neocolonial extractivism and capitalism. Unlike the dominant environmental movement in rich countries, which tends to work through legal and policy channels that assume the beneficence of the state, environmentalism from below often militates against state power. While not averse to putting pressure on the state and elites, environmentalism from below prizes popular autonomy. Originating in the lives of marginalized or subaltern communities and their links to endangered worlds, environmentalism from below is a self-generating and unruly power.[14] It protects the imperiled world by throwing sand in the gears of capital and the state, using direct-action tactics such as blockades and mass demonstrations.

The chapters that follow will explore environmentalism from below in far more detail. We will travel through time, linking past and present struggles against the environmental despoliation of colonialism and capitalism. In addition, we will also travel through space, for environmentalism from below weaves together the majority populations of the Global South with people's movements in the industrialized countries fighting environmental racism and other climate injustices.

The geographical term *Global South* designates power relations in a capitalist world-system that has undergone significant shifts in recent decades. Synonyms could include *poorer nations* or *peripheral countries*, antonyms to the "advanced," core, or imperial nations. We may perhaps see the idea of the *Global South* as imperfect successor to the *Third World* since it skates over the growth of wealthy strata in some poorer nations. I use it throughout this book since I believe that the world-system remains stratified along lines established during the colonial era. Emissions inequality between nations and regions of the world is still stark: the average person living in sub-Saharan Africa produces just 0.6 tons of carbon dioxide each year, while the average US citizen produces 14.5 tons a year.[15] What has shifted in recent decades is emissions inequality *within* nations. While the average inhabitant of a Global North country unequivocally has polluted more than people in the rest of the world, rising inequality within nations means that the average carbon footprint of the world's top 1 percent is more than seventy-five times higher than the bottom 50 percent. Global elites are responsible for a quarter of the growth in carbon emissions during the period from 1990 to 2019.[16] When I use the term *Global South*, I am of course not referring to global elites living in poorer nations but rather to those who are on the sharp edge of ecological breakdown, racist and sexist oppression, and neo-colonialism—the frontline and vulnerable communities of this world. It is these oppressed classes who are rising up most militantly against fossil capitalism's business as usual.[17] It is they who are the most creative and intransigent opponents of a system predicated on the annihilation of nature. It is they who most exemplify environmentalism from below.

Our collective fight for a future must be defined by efforts to build solidarity with and between these diverse people's struggles against planetary ecocide, thereby forging a global movement for ecological and social reconstruction. It was precisely this effort to build transnational solidarities and new strategies of resistance that the gathering of movements in Cochabamba was intended to catalyze.

## Environmentalism from Below in Action

I got an immediate taste of what environmentalism from below looks like as I walked the streets of Cochabamba with members of the US delegation in April 2010. Our remarkable, forty-strong group included climate justice activists, a diverse array of grassroots housing activists, union representatives, media producers, leaders of environmental justice organizations like South Bronx Unite, and green urban planners. As a whole, the group represented the US-based frontline communities and their climate justice allies whom Morales saw as a key alternative to the elite state representatives and corporate hacks dominating COP15 negotiations. We joined environmental activists from across Latin America and other continents, a gathering that demonstrated grassroots global mobilization and solidarity in action.

The counter-model of environmentalism from below that flowered in Cochabamba was influenced by the powerful tradition of popular assemblies and communal decision-making that had developed among leftist movements in Bolivia and other Pink Tide countries in Latin America since the overthrow of dictatorships in those countries.[18] Perhaps the most famous example of such popular mass meetings was the 1996 Encounter for Humanity and Against Neoliberalism, organized by the Zapatistas in Mexico,[19] but radical movements within Bolivia had developed their own similar traditions of collective mobilization outside the state, drawing on Indigenous cultural traditions.[20] A movement known as the *Coordinadora*, which formed during the Cochabamba Water Wars, emerged from and reflected these traditions of horizontal, participatory self-organization. Until the electoral victory of the Movimiento al Socialismo (MAS, or Movement toward Socialism) party in 2005, however, such popular mobilizations had always been separate from state decision-making. Morales was attempting to transform the Bolivian state through radical experiments in popular sovereignty, and the 2010 conference was only the latest example of that effort.

The structure of the conference reflected these traditions of popular assembly and consultation. Each day included testimonial sessions at which conference attendees could speak publicly about key issues relating to their particular struggles for climate justice. The open format of these sessions generated a stream of provocative and heterogeneous testimonials, from the efforts of Cochabamba-based activists to establish community-owned radio and TV stations to a call from a German activist from the Climate Action Network for a global day of direct action to protest the abject failure of the Copenhagen summit. The testimonial sessions in Cochabamba were a kind of pedagogy of the oppressed, a space where activists could share knowledge and experiences with one another to build a stronger, more politically mature and globally aware movement. With simultaneous translation from many languages, the sessions also helped strengthen the feelings of empathy necessary to build solidarity across cultural difference.

As I listened to the testimony of campesinos from across Latin America, anti-extractivist activists from Nigeria, Indigenous militants from the Amazon, and residents of Cochabamba who had fought in the Water Wars, it was clear that the climate emergency is the mother of all crises, one that draws together all the threads of inequality and injustice that have characterized this planet over the previous five centuries or so of colonial and capitalist oppression. One message echoed repeatedly: *Pachamama o Muerte!* The path of life or the path of extinction. The path of life was embodied at the conference through repeated invocations of Pachamama, the earth goddess whose veneration is fundamental to Indigenous Andean environmental justice movements.[21] To respect and honor Pachamama is to recognize the agency of non-human beings as well as the earth itself. The worldview of Indigenous environmental justice movements breaks radically from Western ways of seeing the planet as an inert substance to be dominated and "improved"—an ontology that justified colonization and repeatedly lumped not just other species but also racialized subjects like Native Americans and enslaved Africans with a natural world that colonizers could ruthlessly exploit.

The theoretical underpinnings of Pachamama and complementary concepts such as *buen vivir*, or "the good life," were elucidated at a plenary session on the "Rights of Mother Earth." Among those providing testimony at this plenary was Alberto Acosta, an Ecuadorian economist and ex-president of the Constituent Assembly that drew up the Montecristi Constitution, which established protection for the rights of nature when it took effect in 2008. Acosta situated the Montecristi Constitution as part of a longer historical process that includes the abolition of slavery and women's suffrage. "We must," he stated, "begin to reconstruct our perspective on citizenship to include environmental citizenship, to include biodiversity, to include both human rights and the rights of nature." These ideas were lent support by South African lawyer Cormac Cullinan, author of *Wild Law*, who spoke about how the racially oppressive system of apartheid was based on ideologies of separation and superiority/inferiority.[22] Cullinan argued that the legal system is structured around domination, with Earth treated—as slaves once were—as simple property, shorn of rights. Earth jurisprudence breaks with this tradition, Cullinan argued, foregrounding practices of restorative justice that weave people and planet back together. We have to start thinking in holistic terms, Cullinan said, for we are "beings connected to a community of interrelated beings, bound together by intimacy and love."

How could these rousing ideas of interconnection and respect for the earth be made real in the face of the capitalist system's relentless drive to exploit nature, and the political organization of human affairs into competing nation-states? At the plenary, Mari Margil, executive director of the US-based Center for Democratic and Environmental Rights, explained how the export of US legal statutes concerning the environment is helping to justify enclosure and exploitation of the environmental commons around the world. Just like the laws that legitimated slavery, these legal regimes cannot be reformed, Margil argued, but must be abolished. The laws must be rewritten, as they were in Ecuador and then in Bolivia, so that the rights of nature become part of every country's constitution.[23]

In addition to the testimonial and plenary sessions, the conference also featured seventeen working groups tasked with producing specific radical demands and policy goals. These groups hashed out proposals through public debates that were always engaging, and, at times, also quite acrimonious. Nor was the conference as a whole without dissension. Although the Bolivian government intended the seventeen working groups to be focused on codifying the climate justice movement's demands for the UN Framework Convention on Climate Change, a group of Indigenous activists intent on denouncing the Bolivian government's own record of extraction set up a table outside the conference venue. The dissident Mesa 18 group challenged Morales and the MAS party's commitment to the harmonious ways of being embodied in Indigenous values such as *buen vivir*. They argued that the government was continuing to support extractivist projects that were destroying the lands in which Bolivian Indigenous people themselves lived—contradictions most clearly exemplified in the government's plans to build a highway through a national park (known by the acronym TIPNIS) in the Bolivian Amazon.[24] The TIPNIS conflict subsequently split the Unity Pact of five allied national social movements that had brought Morales to power.

Notwithstanding these tensions, however, the eventual declarations that emerged from the World People's Conference's working groups, statements that were collated into a *People's Agreement* fed back into the UN Framework Convention's process by Bolivian ambassador Pablo Solón, offer an enduringly radical program for global ecological reconstruction. The *People's Agreement of Cochabamba* and the Final Conclusions documents drawn up by the working groups are a template that I will return to repeatedly over the course of this book, as they offer a set of goals determined by the needs and aspirations of movements for environmentalism from below.[25] Much has changed since 2010, but the foundational propositions articulated in Cochabamba remain salient and inspiring. I have found a few key themes particularly essential in framing the chapters that follow.

Foremost among these is the insistence that solving the climate crisis inevitably implies confronting capitalism and colonialism on a planetary scale. For while some ecological reforms may be possible on a local level, green capitalism is unsustainable on a global scale. Capitalism is characterized by a relentless drive to acquire surplus value. It is an economic system that requires incessant growth—and since this growth inherently is ecologically destructive, any "green" capitalist accumulation regime will eventually smash up against the limits of nature. As the *People's Agreement* puts it, "This regime of production and consumption seeks profit without limits, separating human beings from nature and imposing a logic of domination upon nature, transforming everything into commodities: water, earth, the human genome, ancestral cultures, biodiversity, justice, ethics, the rights of peoples, and life itself."[26] Capitalism relentlessly commodifies the world, the *People's Agreement* argues, spawning separation and domination, colonialism and imperialism, in a rush to generate profit that is burning up the planet. A truly sustainable future consequently must move humanity and the planet *beyond* capitalism.

The *People's Agreement* underlines that the logic of capitalism is not simply material but also cultural. It leads, in other words, to what the cultural critic Raymond Williams would have called a "structure of feeling," a condition that legitimates the sundering of people's relationship with nature, as well as an exploitative way of being that instrumentalizes every living thing in the pursuit of feckless and unhinged profit. This regime of accumulation ultimately relies, the *People's Agreement* witheringly observes, not simply on ideological domination but on naked force: "Capitalism requires a powerful military industry for its processes of accumulation and imposition of control over territories and natural resources, suppressing the resistance of the peoples. It is an imperialist system of colonization of the planet." If humanity once faced an opposition between socialism or barbarism, in the words of Rosa Luxemburg,[27] today we are at an even more chilling crossroads: "Humanity confronts a great dilemma," the *People's Agreement* argues, "to continue on the path of capitalism, depredation, and death, or to choose the path of harmony with nature and respect for life."

In place of this headlong rush toward annihilation, the *People's Agreement* argues for what might best be termed *decolonial ecologies*: concrete material and cultural alternatives to the death-dealing colonial matrix of power and "universal" ideas of Western modernity and global capitalism such as development.[28] For the authors of the *Agreement*, it is in the Indigenous ways of life and cultural orientations known as *buen vivir* that such decolonial ecologies are most clearly embodied. Only through an embrace of such values will we avoid planetary ecocide: "In order for there to be balance with nature, there must first be equity among human beings. We propose to the peoples of the world the recovery, revalorization, and strengthening of the knowledge, wisdom, and ancestral practices of Indigenous Peoples, which are affirmed in the thought and practices of 'Living Well,' recognizing Mother Earth as a living being with which we have an indivisible, interdependent, complementary and spiritual relationship."

This emphasis on decolonial ecologies informs the demands made in Cochabamba of the industrialized nations. These demands are centrally informed by the notion of climate debt, a subversive concept that inverts dominant understandings of who owes whom in the world today. Climate debt is predicated on the idea that the Global North is indebted to the South, since it is the rich nations that have polluted the atmosphere with their emissions. The *People's Agreement* frames climate debt in multidimensional terms that include: decolonization of the atmosphere by developed countries through reduction and absorption of their emissions; assumption of the costs and technology transfer needs of developing countries arising from lost development opportunities due to rich nations' historical colonization of the atmosphere; payment of adaptation debts to developing nations struggling to cope with the damages resulting from emissions; and, finally, assuming responsibility for the hundreds of millions of people who will be forced to migrate because of the climate change caused by rich countries. Importantly, the *People's Agreement* puts a concrete figure on these climate reparations, calling for historically polluting nations to devote 6 percent of their gross domestic product to aiding transition in the Global South. This would amount to a historically unprecedented remedy for centuries of colonial rapine, potentially healing massive global inequalities that have intensified as a result of the debt crisis during the neoliberal era.

Such reparations are in the interest of the wealthy nations since, as Naomi Klein has noted, it is the development paths adopted by fast-industrializing countries in the Global South that will determine whether we win or lose the battle against climate chaos.[29] The *People's Agreement* argues adamantly that the peoples of the Global South have a right to develop, "to provide basic services for the entire population and a degree of industrialization which allows the country's economic independence." Yet this development, the *Agreement* stipulates, must not harm the environment. Whatever technology transfer takes place as a part of reparations, it must be appropriate to the needs and cultures of local peoples: "Among the technologies we require are: recycling of waste materials, improvement of traditional techniques with new technologies, access to clean energy sources—solar, wind, and biogas digesters, forms of protection against natural disasters, research into vaccines and medicines for diseases enhanced by climate change, among others."

The *People's Agreement* contains many additional concrete policy suggestions that should inform a program for global ecological reconstruction, but perhaps most important is its overarching insistence not simply that the Global North pay back its climate debt but that the peoples and nations of the Global South "establish policies and strategic lines to confront climate change." These arguments for alternative forms of development, ideas and practices that I am calling *decolonial ecologies*, inform the specific proposals to follow about the possible shape of the new world we must build. Key to the proposals, and implicit in everything that unfolded in Cochabamba, is one basic question: How can we decolonize the Green New Deal?

## What Happened to the Global Green New Deal?

The history of efforts to build the global Green New Deal is important to understand since it is essentially a chronicle of the limits and contradictions of environmental politics in the world's core capitalist nations. It must first be said that the Green New Deal was never a singular "deal": it was a congeries of proposals that evolved over time. It was also a rallying cry, a counter-hegemonic project that sought to win popular support for environmental measures through the promise of green jobs. Over the last decade and a half, proposals were put forward in various countries by progressive political parties, academics, and even grassroots environmental movements who often had very different understandings of what was driving ecocide, but who nevertheless saw an opportunity to simultaneously tackle twin calamities of the climate emergency and the crisis of capitalism.

The Green New Deal is also a tendency within more liberal or progressive sectors of capitalism, who can be called "eco-modernists," and who support policies intended to mitigate carbon emissions—but always within a capitalist framework. This means that the Green New Deal needs to be seen as a plural, fragmented, piecemeal, and deeply contradictory set of policies put in place as the scientific evidence of the climate emergency has become irrefutable. The majority of these policies have pushed for policy reforms in areas such as the transition to renewable energy that seek to rebuild infrastructures in order to keep the capitalist system purring along like normal. Everything must change, in other words, so that things can remain pretty much the same—not exactly a viable proposition over the long term.

Lastly, despite gestures toward internationalism, the history of Green New Deal is one of reform within nation-states who see themselves as locked in the kind of zero-sum global rivalry that defines inter-imperial competition.[30] As Harpreet Kaur Paul and Dalia Gebrial put it, "The Green New Deal has been trapped in national imaginations."[31] As a result, while the effort to mitigate carbon emissions that is at the heart of Green New Deal would obviously benefit the planet as a whole, little if any thought is given to Global South nations within the vast majority of Green New Deal proposals.

In sum, the Green New Deal must be decolonized.

Green New Deal discourse can be dated back most decisively to 2007, when, in the context of the Great Recession, British economists Ann Pettifor and Richard Murphy and a group of colleagues founded the Green New Deal Group, arguing that "the global economy is facing a 'triple crunch' . . . a combination of a credit-fueled financial crisis, accelerating climate change, and soaring energy prices underpinned by an encroaching peak in oil production."[32] To cope with this "triple crunch," the Green New Deal Group proposed a suite of policies that would entail "re-regulating finance and taxation plus a huge transformational programme aimed at substantially reducing the use of fossil fuels and in the process tackling the unemployment and decline in demand caused by the credit crunch."[33]

The Green New Deal Group's proposal resonated widely, and within a year the European Green Party had proposed a Green New Deal for Europe, and the United Nations Environmental Program had published its own blueprint for what it called a *Global Green New Deal*.[34] All of these proposals followed the lead of Britain's Green New Deal Group in outlining measures for an ecological overhaul that simultaneously aimed to deal with the Great Recession by spurring economic growth. They were, in sum, a form of Environmental Keynesianism: government spending intended to offset the cyclical downturn of capitalism, increasing demand and employment. It's just that state spending, in this case, was directed toward a "green" overhaul of the infrastructures of fossil capitalism.[35]

Despite the quick spread of conversations about a Green New Deal in Europe and the US during the Great Recession, ambitious programs of ecological reconstruction never really got off the ground. In Britain, a general election in spring 2010 produced a coalition government led by the Tories. Almost immediately they began implementing a program of draconian austerity that doomed any discussion of a state-led revival of the economy.[36] Talk of a Global Green New Deal from organizations like the United Nations also dried up quickly as the political winds shifted. Things looked more promising for significant environmental reconstruction in the US. The Supreme Court had ruled in the landmark *Massachusetts v. Environmental Protection Agency* case of 2007 that the federal government had the right to regulate greenhouse gas emissions through the Clean Air Act. The victory of the relative outsider presidential candidate Barack Obama seemed to promise transformational change on the environmental front. As an indication of promising policies to come, Van Jones, who was to become Obama's "green jobs czar," published a manifesto in October 2008 calling for an "eco-populism" grounded in "building a New Deal coalition for the new century" that would include labor unions, environmentalists, students, faith groups, and social justice activists.[37]

Yet rather than imposing emissions cuts using the power of the EPA, Obama decided to negotiate with congressional Republicans. Unsurprisingly, no policies were passed, and the public was subjected to a drawn-out spectacle of humiliating defeat that seriously deflated the "hope" for the future that Obama had conjured up in his campaign. Despite this failure, many pundits writing about the prospects for the Green New Deal after the election of Joe Biden in 2020 argued that Obama's policies offered a template for successful climate action a decade later.[38] To support this surprising claim, they pointed to the $90 billion that Obama poured into renewable energy, batteries, smart grid innovations, energy efficiency, and other green policies. Given the neoliberal orientation of the Obama administration, this funding for energy transition took the form of public benefits for private renewable companies (and big banks) rather than the creation or expansion of public power.[39] In addition, the impressive growth of renewable energy in the US would not have taken place without the efforts of the German and Chinese governments, who massively expanded production of solar and wind technologies through state subsidies that subsequently helped to reduce the price of the technologies for US consumers.

Another reason that the Obama Green New Deal is often forgotten is that it unfolded alongside and was more than offset by a historic surge in extractivism that turned the US into the world's premier producer of oil and gas. Under the Obama administration, the Federal Reserve essentially bankrolled the development of a whole new regime of fossil capitalism by locking in low interest rates that allowed fracking companies to borrow trillions of dollars. In 2000, only 2 percent of oil and gas wells drilled in the US were for fracking; today, 69 percent of wells are fracked.[40] Fracking firms lost astounding sums of money and went bust in droves for much of the last decade, but that didn't stop the fracking boom. As David Wallace-Wells put it in 2022, "Fracking has been, for nearly all of its history, a money-losing boondoggle."[41] The massive public subsidies that spawned the fracking boom could have and should have gone to support the build-out of publicly controlled renewable energy.

The Biden administration's Green New Deal seemed set to follow a very similar trajectory to Obama's GND. Biden came into office siphoning the political energies unleashed by the insurgent climate Left, which had made a public splash when newly elected representative Alexandria Ocasio-Cortez and members of the Sunrise Movement occupied the office of Democratic Party leader Nancy Pelosi. AOC, a democratic socialist (a rare beast in the US political jungle), brought both radical urgency and a sense of historical weight to her fight for the GND, calling it "the Great Society, the moon shot, the civil rights movement of our generation." Biden's climate plan responded to this climate insurgency by promising to spend $2 trillion to completely decarbonize the US economy by 2035.[42]

Once in office, however, Biden appeared to follow the same playbook as Obama, spending nearly two years courting right-wing members of his party like West Virginia senator Joe Manchin, a millionaire coal baron who took more money from fossil fuel companies after Biden came to power than any other senator.[43] These negotiations nearly proved to be just as fruitless as Obama's efforts to win over climate change–denying corporate groups, until a dramatic turnaround late in the summer of 2022 led to the passage of the uncharismatically titled Inflation Reduction Act (IRA). Really a climate change law, the IRA set aside $369 billion in funding for clean energy and electric vehicle tax breaks, pollution reduction (including methane leaks), cleanups in environmental justice communities, and domestic manufacturing of batteries and solar panels.[44] The IRA also contained funding for pipelines and offshore oil and gas drilling leases, a bribe to Manchin the fossil man. As Colette Pichon Battle from Taproot Earth Vision (formerly the Gulf Coast Center for Law & Policy) put it, "Once again, the only climate proposal on the table requires that the communities of the Gulf South bear the disproportionate cost of national interests bending a knee to dirty energy—furthering the debt this country owes to the South."[45] Nevertheless, champions of the measure point to assessments of the likely impact of the IRA that suggest it will reduce US carbon emissions 40 percent from 2005 levels by 2030, putting the nation on track to meet its Paris climate agreement targets.

Reactions to the IRA's passage were hyperbolic among the liberal commentariat, with columnist Paul Krugman penning a piece headlined, "Did the Democrats Just Save Civilization?"[46] The IRA is certainly better than nothing, but it is worth remembering that scientific analysis of the climate pledges made at Paris in 2015 showed that they are not nearly enough, and, even if adhered to, will put the world on track to nearly 3°C of warming.[47] As Jean Su of the Center for Biological Diversity put it, the IRA "doesn't get us anywhere near where we need to go."[48]

In addition, six months after passing the IRA, the Biden administration gave the green light to ConocoPhillips's massive Willow oil development on the remote tundra of Alaska's northern Arctic coast.[49] Breaking Biden's campaign promise of "no more drilling on federal lands, period," the Willow development will extract 600 million barrels of oil over the next thirty years. This one project will produce as much as Belgium's total carbon emissions over that 30-year stretch. And it is just one among nearly seven thousand drilling leases that Biden approved during his first twenty-five months, making his administration responsible for an ecocidal fossil fuel binge outpacing that of the Trump administration. So much for the US's first "climate president."

With Green New Deal policies—such as they are—limited exclusively to the scale of (core) nation-states, there is little space to reckon with much less to counter the predatory impact of a global capitalist system based on intensifying extraction of resources from nations in the Global South. Environmental justice activist Asad Rehman worries that dominant Green New Deal policies will amount to a new green colonialism. As Rehman argues, "Britain is planning to go green through a new phase of resource and wealth extraction of countries in the global south."[50] He expresses particular concern that the imperative to shift to renewable energy will lead to a new round of extraction of minerals such as cobalt, lithium, and nickel that are critical to clean power infrastructure.

Rehman is not alone in warning of the possible advent of a new green colonialism. The making and expansion of global capitalism has always been inextricably linked to colonial relations and the forced extraction of large quantities of natural resources from the Global South. The so-called enclaves where extraction takes place experience few of the financial benefits and all of the environmental and social destruction. These exploitative relations are five hundred years old, and the legitimacy of the state has always been tied to large mining and infrastructure projects. But in recent decades, self-avowedly socialist governments in parts of the Global South have sought to legitimate continued extraction by increasing the portion of profits captured by the local state, which redistributes those revenues through social programs. Yet, as I will explore in more detail in a subsequent chapter, people in mining regions continue to suffer negative social and ecological consequences. This tension around so-called neo-extractivism is what was behind the Mesa 18 protests that I witnessed in Cochabamba: Indigenous residents of the rainforest where oil drilling was set to take place refused to surrender their land, even to a state headed by an Indigenous president. These conflicts are likely to intensify as the quantities of minerals needed for Green New Deals in the Global North increase dramatically.

In a reflection of mounting concern about green colonialism, frontline defenders from across Latin America, meeting in Chile shortly before the scheduled UN Climate Summit in autumn 2019, issued a declaration that stated: "Hidden behind the discourse of 'energy transition' is a program of economic growth for the Global North which threatens to exponentially increase sacrifice zones under the auspices of guaranteeing the supply of minerals for so-called 'green technologies.' This will come at the cost of the exploitation of our territories and communities, all while intensifying the ecological crisis."[51] What, these activists ask, is to prevent a Green New Deal, especially one framed around boosting employment and demand in the Global North, from intensifying the plunder of poorer nations' resources?[52] These concerns were no doubt intensified by Evo Morales's allegation that he was ousted from power in response to his efforts to nationalize Bolivia's lithium supplies, an argument that appears to be unsubstantiated but that many nevertheless found easy to believe given the US's history in the region and around the world.[53]

It is undeniable, however, that energy transition is set to dramatically intensify extractivism. An average electric vehicle requires six times the amount of minerals as a conventional car, according to the International Energy Agency, including precisely the ones that Asad Rehman mentions in his warning about green colonialism.[54] An onshore wind turbine requires nine times more mineral resources than a gas-fired power plant. The World Bank estimates that demand for minerals could increase tenfold by 2050.[55] Some minerals are already seeing demand spikes: global lithium production shot up from 18,000 tons in 2009 to 86,000 tons in 2019. Some analysts have warned that the massive increase in demand for minerals will outstrip supply.[56]

But this is largely a political rather than a material problem. There is no shortage of lithium in the earth's crust: in 2020, 86 million tons of lithium resources were identified, equivalent to the needs of 8.6 billion electric vehicles, according to the US Geological Survey.[57] As demand increases, more supplies tend to be located. At issue is the fact that many of these minerals are not located in European Union nations and the US, and that the core countries have been slow to develop supply chains for these minerals. China has not. It controls the processing of key energy-transition minerals, including an estimated 80 percent of the lithium battery chemical market.[58] Hence the efforts to expand domestic production of renewable energy technologies in recent US climate legislation. The growing need for minerals and the geopolitical concerns about their availability help explain why the Green New Deal is not only catalyzing a new green colonialism but also sparking new inter-imperial tension and competition.

It does not have to be this way. The Green New Deal should not be designed around the precept that we must change everything in order to keep the capitalist system and the social relations it generates cemented in place. But that's how dominant efforts approach energy transition. Take transportation: Enormous sums are currently being thrown at US residents in an effort to get them out of internal combustion cars and into electric vehicles. Setting aside the fact that EVs today are mainly powered by electricity derived from burning methane, the assumption behind these policies is that the majority of Americans would sooner see the end of the world than take a public bus or a subway—let alone ride a bike. While it's true that infrastructure has a certain material and cultural heft, it's not really set in stone. If the original New Deal in the US helped produce a landscape of superhighways, racially segregated suburbs, oppressive gender regimes, and shopping malls, generating what one historian has called "a consumers' republic," this took place when there seemed to be no contradiction between Keynesianism's traditional emphasis on stimulating economic expansion and ecological limits to growth.[59] Today that infrastructure is in decay and in grave need of replacement. What it gets replaced with is a political decision.

Instead of boosting growth in general, a globally aware, intersectional, and decolonial Green New Deal could stimulate expenditure and employment in sectors such as renewable energy, care work, and public transportation while simultaneously constraining and contracting the expansion of extraction and feckless consumption. Naomi Klein argues, for example, that "any credible Green New Deal needs a concrete plan for ensuring that the salaries from all the good green jobs it creates aren't immediately poured into high-consumer lifestyles that inadvertently end up increasing emissions."[60] For Klein, the transition must have "hard limits on extraction that simultaneously create new opportunities for people to improve quality of life outside the endless consumption cycle."[61]

Absent such hard limits on extraction, the Green New Deal will perpetuate what the German cultural critics Markus Wissen and Ulrich Brand call "the imperial mode of living": the neocolonial appropriation of nature inherent in patterns of mass consumption in wealthy nations and among elites in developing countries.[62] The flip side of the precepts of *buen vivir* that I witnessed in Cochabamba, the imperial mode of living assumes unfettered access to unlimited resources—from atmospheric space for carbon emissions, to raw materials, cheap labor, and unlimited sinks for all the pollution produced from mass consumption—that only a small segment of the global population can enjoy. Although the imperial mode of living is woven into everyday life in the wealthy nations to the point that it has become largely invisible to the affluent, it is dependent on maintaining access to natural resources through political and legal coercion, or, when these means fail, through the application of brutal military force.

This must end. So, yes, a truly decolonial Green New Deal *will* ground private jets, it *will* heavily tax luxury penthouse apartments, it *will* defund the police and the military, and it *may* even make hamburgers more expensive. But it will also fund public parks and playgrounds, public art and public transit, public housing, libraries, and green schools. In short, it will be characterized by what the journalist George Monbiot calls "private sufficiency and public luxury."[63]

Yet as long as Green New Deal advocates overlook established ownership patterns, they will leave existing class relations and the forms of exploitation they entail intact.[64] Klein's allusion to the problem of "throwaway crap from China"—shorthand for Keynesian policies that create jobs but also encourage overconsumption—raises the problem of the current international division of labor, in which significant quantities of production have been offshored to developing nations by corporations such as Apple and Nike. This division of labor has helped perpetuate the illusion of decoupling: the idea that production has grown more efficient and hence greener in "advanced" nations such as the US. Embraced by both mainstream economists such as Nicholas Stern and Silicon Valley tech gurus,[65] the notion of decoupling ignores the fact that dematerialization in rich countries such as the US relies on macro-scale super-materialization in other nations.[66] Indeed, recent history offers good cause to be concerned about a new green colonialism. Global extraction and consumption of raw materials in general—including fossil fuels—increased eightfold during the period from 1900 to 2005.[67] The increasing rate of extraction has been particularly pronounced during the era of neoliberal globalization, with global consumption growing by 93.4 percent between 1980 and 2009.[68]

This consumption is hardly evenly distributed around the world: the current environmental crises owe almost entirely to a group of around one hundred corporations and a small stratum of obscenely wealthy people. This should not be so surprising: as the anthropologist Jason Hickel argues, the last fifty years should be known as the "Great Divergence" since, while global gross domestic product has grown steadily, only about 5 percent of that income goes to the bottom 60 percent of humanity.[69] This followed a brief interregnum from 1950 to 1980, during which decolonization and the efforts of the Non-Aligned Movement to establish a New International Economic Order increased economic sovereignty for postcolonial nations and narrowed the economic gap between rich nations and their former colonies, generating higher wages, improved access to land for peasant farmers, and a greater share of national income for workers and the poor across the Global South. Today, 50 percent of the resources that high-income nations consume are appropriated from the Global South, suggesting that neocolonial extractivism has intensified markedly over the last half century. While the unfettered, brutal extraction of so-called natural resources is as old as colonialism, today's global system of hyper-capitalism has already transgressed key planetary ecological boundaries.

As the activists who assembled in Cochabamba argued, this is the path to planetary death. The Indigenous Environmental Network puts it clearly: "An economy based on extracting from a finite system faster than the capacity of the Earth to regenerate will eventually come to an end."[70] A recent review of hundreds of scientific studies showcases the direct link over the last half century (during the so-called Great Acceleration) between economic growth, energy use, and carbon emissions.[71] While the original New Deal helped spur this uptick, the Green New Deal must wind it down.

Perhaps retaining the term *Green New Deal* is a mistake, since the New Deal itself was intended to save the US's capitalist system, excluded most people of color and women within the US, and paved the way for an economy based on permanent imperial warfare.[72] Whatever we call it, though, ecological reconstruction must be anti-capitalist, feminist, and decolonial—which means that dismantling the imperial war machine must be a primary component of climate justice. It means that ecological reconstruction must include adequate funding for the work of social reproduction, labor that has been shouldered overwhelmingly by women and the poor. People must also just work less, getting off the treadmill of unnecessary consumption and figuring out how to reknit frayed social relations.

It also means that ecological reconstruction must unfold within the kind of reparative framework articulated in the Cochabamba *People's Agreement*, one that places primary responsibility on historic emitters in the Global North to shoulder their fair share of the fight for a sustainable world. It implies not just scaling up ambition to decarbonize society as rapidly as possible, but also ensuring that Global South countries are provided with the economic resources, knowledge, and rights necessary to deploy appropriate clean technology themselves. And this transformation cannot stop at energy. It must also dismantle dominant trade regimes and the institutions that enforce them (such as the World Trade Organization) in order to permit countries to regulate and abolish environmentally destructive extractivism. We need to rethink how we grow our food, how we build our cities, and who gets to control the planet's forests and other lands. In other words, as we will see in the coming chapters, we need a fundamental reconstruction of the global economy and society that draws on the knowledge and activism of movements for environmentalism from below.

## A People's Plan for Ecological Reconstruction

Liberation comes from below. I felt sure of this as I boarded the plane to return from Cochabamba to New York. The exhilarating experiences of environmentalism from below that I had had in Cochabamba left me feeling hopeful about the prospect of building the global movement for climate justice into a force capable of turning the tide of climate denialism and ecological devastation. The time seemed ripe to construct a people's plan for ecological reconstruction.

But then I saw the *Deepwater Horizon* in flames. We'd touched down in Miami for a connecting flight to New York. As my comrades and I walked to our gate, a television monitor played the horrific scenes over and over: flames gushing out of the collapsing superstructure of the oil rig; minuscule-looking rescue boats spraying thin streams of water uselessly onto the billowing flames; oil bleeding in huge gouts into the sea; a long stain of pollution spreading out across the waters of the Gulf of Mexico. The guts of the planet seemed to be pouring out, spewing such a massive slick of illness and pollution that no place on Earth could be left clean or whole. My friends and I turned to one another with shock, pain, and despair. Revulsion and anger knuckled people's faces. Never in my life had I experienced a moment of such great hope foreclosed with such rapidity and apparent finality.

But to despair at an event such as the explosion of the *Deepwater Horizon*—and at the political corruption that allowed it to occur—is a luxury that no one can afford at this point in world history. We have not yet won the radical demands that were articulated in the Cochabamba *People's Agreement*, but they remain an inspiration and a lodestar for movements fighting for global ecological reconstruction. Participating in the climate justice movement in the years since Cochabamba has shown me that people will continue to fight tenaciously, will constantly strive to produce creative and daring alternatives to the dominant arrangements, no matter how high the odds are stacked against them. This is because for most people around the world, there is no alternative but to fight.

This book offers a chronicle of struggles for environmentalism from below. Sometimes this involves analysis of coordinated transnational popular movements, but often it includes accounts of dispersed groups of people and communities whose combined efforts are helping to move forward the fight against planetary ecocide. Either way, in order to understand the shape and political strategies of these movements, we must understand the forces arrayed against them. For this reason, each chapter explores not just movements for environmentalism from below but a far broader landscape related to critical areas of environmental crisis today.

I begin with agriculture and the global food system. To resonate in the Global South, any proposal for global ecological reconstruction must commence with agriculture, the sector in which the majority of the world's people work. Of all its disastrous impacts, climate change is going to affect farming—and our ability to grow enough food to meet humanity's needs—the most. Farming communities in the Global South are already struggling to stay in place, pinched between elite landlords and a global system of industrial agriculture that is making subsistence more and more difficult. Climate change will dramatically exacerbate these challenges, placing immense pressure on women in particular, who do the vast majority of agricultural labor in poor countries. Over time, temperatures around the world will rise, droughts will increase, crop yields are likely to decline, and livestock will face starvation. Small farmers are increasingly being peddled genetically modified and purportedly drought-resistant seeds by global agricultural conglomerates, but such seeds are often not well adapted to local conditions, and their purchase puts small farmers deeper in debt.

Rather than such poison pills, farmers need access to more land and water, which means there needs to be genuine land reform across much of the Global South. In addition, farmers need advice about sustainable and low-cost agroecological farming techniques, and subsidies to cope when their crops fail as a result of climate change. Finally, they also need access to climate jobs, building and maintaining farms of wind turbines and solar panels to generate power. I explore the potential of sustainable agroecology as practiced by peasants not only to provide an alternative to industrial agriculture but also to absorb massive amounts of carbon to boot.

The next chapter of the book looks at cities in the Global South. The human condition is now an urban one, as more than 50 percent of people live in cities. Soon that urbanized proportion will be far higher. The vast majority of urbanization this century, however, will take place in poor nations, producing many more megacities like Lagos, Dhaka, and São Paulo, as well as thousands of cities with smaller but nevertheless cumulatively world-altering populations. Policies of the neoliberal era have already produced what Mike Davis called a "planet of slums," whose inhabitants now face searing heat, water stress, rising sea levels, and mega-storms as climate chaos intensifies. People in these cities need resources for adaptation as well as decent, dependable jobs. Most of all, these cities must avoid following the US pattern of automobile-based, fossil-fueled suburban sprawl. This means that people need access to affordable solar panels and other forms of renewable infrastructure. Local municipalities need to upgrade existing buildings and construct new zero-emissions housing. Cheap and accessible systems of public transportation such as Bus Rapid Transit systems, which have already been shown to be successful in developing nations, must be constructed on a mass scale. And this green technology should be manufactured locally to counter the exclusion of people from the formal economy, who are then reduced to the status of "surplus humanity."

This chapter explores the evolving character of urban climate insurgency, or the responses of global majorities, today's equivalent of the people whom Frantz Fanon called "the wretched of the earth," to the increasingly extreme conditions of the contemporary urban order. I argue that urban climate insurgency includes the resistance of slum dwellers to eviction to the urban periphery, which not only severs people from their social networks, but also generates ecologically unsustainable forms of urbanization. The chapter looks at examples of popular slum-upgrading projects that help secure people's land tenure as well as giving them access to good homes capable of withstanding the challenging climate conditions increasingly affecting the world's cities. These beneficial transformations of the social and ecological fabric of today's extreme cities are particularly effective when carried out through the projects of mutual aid that characterize housing collectives. In such a context of popular mobilization, insurgent urban climate movements can help shift the broader political culture of cities to support grassroots sustainability.

Greening the built infrastructure of the cities of the Global South will have little impact unless their denizens and fellow citizens in rural areas get access to clean power. Struggles over access to electricity in Global South cities as well as in the countryside are interwoven with fights around fossil capitalism. In many Global South nations, coal is playing an increasingly decisive role in the energy mix, notwithstanding official rhetoric about energy transition. The implications are stark, since these societies are undergoing transitions from predominantly biomass-based to fossil fuel–based systems, locking in forms of fossil development at the precise time when the world needs to cease burning fossil fuels. Will the masses in the Global South go along with fossil developmentalism, or will they fight for alternative forms of power? What should energy transition look like in the context of economically constrained Global South nations?

This chapter explores anti-extractivist campaigns that constitute what I term *Global Blockadia*. This transnational movement is succeeding: over a quarter of fossil fuel and low-carbon energy projects documented by the Global Atlas of Environmental Justice project were canceled, suspended, or delayed when met with organized resistance.[73] Given increasing demand for energy in much of the Global South, Blockadia's rejection of fossil infrastructure needs to be accompanied by viable proposals for a just transition to renewable power. In many Global South countries, terms like *just transition* are regarded with significant skepticism given sky-high rates of unemployment and the neoliberal orientation of renewables deployment to date. But energy transition need not involve a new round of corrupt and demoralizing postcolonial politics. Another world based on energy democracy is possible. Publicly owned and democratically managed renewables must be part of a just transition.

Alongside the climate crisis, we are living through the Sixth Extinction, with levels of biodiversity around the world crashing at alarming rates. The dominant response from established conservation organizations has been to throw into overdrive their efforts to enclose more land in so-called Protected Areas, places that are off-limits to people—including the Indigenous people who have often stewarded such lands for millennia. As the conservation industry encloses more and more of the lands collectively owned by Indigenous people, peasants, and pastoralists around the world, the dispossessed are increasingly rising up in opposition. Their cry: Decolonize conservation!

For the growing movement opposing what they call the "conservation industry," the best way to fight against biodiversity loss and climate change is to respect the land rights of Indigenous peoples and other local communities, who protect 80 percent of the world's biodiversity. This chapter explores the context for this campaign to decolonize conservation, looking at the emergence of today's "fortress conservation" policies in European efforts to mitigate the destruction of ecosystems in postcolonial territories such as the Caribbean and India through policies that create millions of conservation refugees. After surveying the colonial origins of conservation, the chapter discusses the orientation of the conservation industry in recent decades, which has primarily involved an avid embrace of the market-oriented approaches known as "Nature-based Solutions." The plan to expand Protected Areas to cover 30 percent of the earth's terrestrial surface by 2030—promoted with the catchy tagline "30x30"—is only the most prominent of the Nature-based Solutions currently being promoted by the conservation industry. After surveying and critiquing these dominant approaches to conservation, the chapter closes with a discussion of the alternatives proposed by the movement to decolonize conservation.

Finally, the concluding chapter of the book draws on the insistence of the Cochabamba *People's Agreement* that climate migration is a human right. Migration is a form of climate adaptation, and, I argue, one that makes perfect sense given the refusal of global elites to slash carbon emissions. This chapter explores efforts to cope with significant waves of climate-related migration both within specific Global South nations and within entire regions of the Global South. Contrary to much anti-immigrant discourse in Global North countries, most climate-related migration takes place within the Global South. I argue that the industrialized nations which have colonized the atmosphere must pay their debts by aiding people in the Global South to adapt to the climate crisis, but also by giving harbor to climate-displaced people. Affirming the right to mobility involves challenging and transforming the legal apparatus governing refugee rights around the globe. And it also means dismantling the border-security complex. In place of the insular and xenophobic trends that are giving birth to horrifying new forms of eco-fascism, I propose new forms of global solidarity and what I call "disaster communism," inspired by movements for environmentalism from below.

As I think about what this global solidarity might look like, my mind goes back to an experience I had shortly after arriving in Cochabamba for the World People's Conference. While I waited to enter the conference venue with my friends Tanya Fields, of the Bronx-based Black Feminist Project, and Byron Silva, an Ecuadorian-born union organizer also based in New York, an Aymara woman named Rosa Graciela Quiroga introduced herself and asked if she could take her photo with Tanya, whose long dreadlocks she admired. While they snapped a photo, Byron reminded me that only ten years ago, bowler-hatted and pollera-skirted Indigenous women like Rosa—who were ubiquitous in the university grounds and lecture halls where the conference was being held—would not have been allowed onto the campus. After the photo was taken, Rosa welcomed Tanya in the name of the Union of Multi-Active Women of Bolivia. The two of them exchanged a hug, and Rosa told Tanya that she considered herself Tanya's sister. In that brief but emotion-laden exchange, a world of radical transformation seemed to be dawning.

# CHAPTER 1

# DECOLONIZING FOOD

O n January 26, 2021, the day India celebrates its anniversary as a constitutional republic, farmers driving thousands of tractors joined hundreds of thousands of compatriots on foot in a march from protest encampments on the fringes of Delhi toward the center of the city.[1] After months spent camping in the cold and rain on the outskirts of Delhi, the farmers were fed up. The protest encampments were set up in late 2020 in an effort to get the ruling Bharatiya Janata Party (BJP) to repeal a trio of laws intended to privatize the Indian agricultural sector. The government of Prime Minister Narendra Modi refused to budge, so the farmers decided to take their protest into the center of the country's capital city. Government efforts to stop the protest using legal injunctions had failed, but the leaders of farmer unions had agreed to keep the protesters on police-approved routes through the city outskirts—ensuring that the protest would not disrupt official anniversary celebrations.

The farmers were having none of it. At the city's border with the village of Ghazipur, site of one of the farmer encampments, tractors pushed aside a shipping container placed on the road by the Delhi police. While Modi officiated at a military parade, protesters broke away from approved protest routes and clashed with police armed with bamboo batons. Elsewhere, police responded with thick clouds of tear gas and baton charges as farmers tried to march off the capital's ring road and push toward the city center. Delhi police commanders placed officers with assault rifles across key routes into the city, but the farmers refused to be kept out of the symbolic heart of the city. By noon, they had breached the Red Fort, the iconic palace in the center of Delhi that once served as the residence of India's Mughal emperors.

Mobilizing farmers and their allies by the hundreds of thousands, the protests in India were one of the biggest civil society mobilizations in the world during the convulsive year of the COVID-19 pandemic and global Black Lives Matter uprisings. The overt intent of the farmer demonstrations was to agitate for repeal of a set of laws rammed through the Indian parliament in September 2020 by the right-wing BJP regime. These laws aim to dismantle the Independence-era system of government regulation that assures farmers are paid a "minimum support price" (MSP) for their crops, thereby protecting Indian farmers, as well as the general public, from the vicissitudes of the free market, both local and global.[2] Although the government markets only buy one-third of the crops produced in India, the MSP acts as a benchmark figure that shapes negotiations for prices throughout the agricultural sector. Under the new laws, farmers would be allowed to sell their produce directly to private buyers rather than at the state-regulated marketplace, to enter into legal contracts with private companies for their crops before harvest, and to hoard grain until prices increase. Activist farmers were outraged by the swift passage of the laws, arguing that they would allow large corporations to displace the small traders who currently dominate the government-operated marketplaces, drastically shifting the balance of power and curtailing farmers' capacity to negotiate fair prices for their crops. Since over 50 percent of the Indian population works in the agricultural sector, the privatization of agricultural markets has stark implications for the country.

Behind this battle over the privatization of agriculture lies the climate emergency. In India over the last half century, extreme weather events, particularly torrential downpours, have tripled in number. Rainfall has grown less frequent, but when rains come they are epic and frequently destroy all the crops in a region. Since roughly half of India's farmers are too poor to afford irrigation and therefore depend on rain-fed agriculture, increasing weather extremes translate into intense economic vulnerability. But even more well-off farmers are in crisis, since they typically attempt to cope with weather extremes by increasing the use of expensive inputs like chemical fertilizers, pesticides, and irrigation, which degrades the quality of soil and lowers groundwater levels.[3] The upshot is decreasing productivity and increasing debt levels, as farmers spend more to grow less. Their efforts to switch to less risky crops ironically lead to overproduction and the crashing of prices for both major and minor crops. As a result of this vicious economic cycle, Indian agriculture increasingly exhibits the same instability and extremes as the weather conditions imposed by the climate emergency.

Caught in this climate vice, farmers have demanded increases in the MSP set by the government, but the Modi regime has instead decided that it is unsustainable for the government to continue to subsidize an increasingly uncompetitive sector. Cut loose by the state, farmers face impossible choices. Increasing numbers have been taking their own lives by drinking pesticides. The insurgency in Delhi is a defiant antithesis to such gestures of hopelessness.

Farmers in India are not the only ones rising up. The protests outside Delhi need to be seen in the context of a deep crisis in the global food system. The world is in the grip of a hunger pandemic. According to the latest report from the United Nations, global hunger rose to 828 million people in 2021, an increase of 46 million since 2020 and 150 million since the outbreak of the COVID-19 pandemic.[4] Worse still, 2.3 billion people—nearly a third of humans alive today—faced moderate or severe food insecurity. Acute food insecurity means that a person's inability to consume adequate food puts their lives in immediate danger. Even in nominally rich nations such as the US and UK, millions are struggling to put food on the table and to access meager government food aid: in 2021, for example, 33.8 million people, or over 10 percent of the US population, were afflicted by food insecurity.[5] Hunger also affects Black and Latinx households disproportionately, with food insecurity rates in 2021 at triple and double the rate of white households, respectively.

The devastating impact of the food crisis is most evident in the toll it is taking on women and girls, who make up the majority of the world's food producers but also suffer disproportionately from hunger. Constituting 43 percent of the agricultural workforce across the Global South in general, women and girls produce nearly 70 percent of food in sub-Saharan Africa, to give one example.[6] But their central role in global agriculture does not guarantee their access to food. Indeed, women and girls currently make up 60 percent of those facing chronic hunger around the world.

The hunger pandemic is also a compound crisis: according to the World Food Programme, political instability, environmental deterioration, and climate crisis are the main factors driving increases in food insecurity over the 2010s.[7] Yet these factors are often related; environmental crisis, for instance, sparks social and political conflict. Exacerbating these chronic conditions, lockdowns triggered by the pandemic disrupted globe-spanning food chains.[8] Food became more expensive as far-flung food chains crashed. Meanwhile, the global economic recession caused by the pandemic produced lower incomes. The upshot was that vulnerable communities and countries saw their access to food collapse.[9] Russia's invasion of Ukraine, one of the world's biggest grain producers, drove global food prices to record highs in 2022, worsening already dire levels of hunger generated by the pandemic, droughts, inflation, and other regional wars.[10]

Things were not supposed to go this way. For decades, elite policymakers argued that creating a global market for agriculture would ensure that affordable food would flow to those most in need. In the face of the intensifying hunger crisis, some continued to embrace free-market doctrines, declaring that opening global markets further would, as the *Economist* put it, "keep things cornucopious."[11] Yet, as we will see in this chapter, many activists and researchers have, also for decades, challenged this free-market dogma. The COVID-19 pandemic made it clear that the food crisis is a product of the increasing vulnerability of global supply chains to contagious failures of one kind or another.[12] Fragile global supply chains subject to the highly volatile forces of financialization are combining with the increasingly precarious nature of work around the world to provoke unprecedented levels of misery and starvation. This vulnerability is magnified by the extreme weather patterns the climate crisis is generating. The increasing precariousness of the global food system should provoke comprehensive policy reforms that strengthen the resilience of the food system, according to the Committee on World Food Security's High-Level Panel of Experts on Food Security and Nutrition. But, as we shall see, these kinds of marginal tweaks are not adequate to deal with the profound contradictions of the global food system.

The depth of the crisis is indexed by the global collapse of biodiversity. In the spring of 2019, the world learned definitively that species extinction rates are accelerating and that the natural world is undergoing a jaw-dropping, dangerous collapse. In the month of May, as flowers burst into bloom and migrating birds returned to trees and fields in the Northern Hemisphere, the Intergovernmental Science-Policy Platform on Biodiversity and Ecosystem Services (IPBES) published a global assessment drawing on an extraordinarily wide range of research. The report synthesized findings that had heretofore examined the crisis of biodiversity in relatively isolated and fragmented spheres. The conclusions were shocking, not simply for the planet's animals and plants but also for the human populations that depend on them. IPBES Chair Sir Robert Watson summarized the landmark report in the following stark terms: "The health of ecosystems on which we and all other species depend is deteriorating more rapidly than ever. We are eroding the very foundations of our economies, livelihoods, food security, health, and quality of life worldwide."[13]

In the most comprehensive assessment of its kind, the IPBES report found that around one million animal and plant species are now threatened with extinction, many within decades. Extinction rates are at their highest in human history, and the problem is not isolated to the total disappearance of particular species. According to the report, "the average abundance of native species in most major land-based habitats has fallen by at least 20%, mostly since 1900. More than 40% of amphibian species, almost 33% of reef-forming corals, and more than a third of all marine mammals are threatened."[14] In an alarming summary of the panel's findings, report co-chair Professor Josef Settele said, "Ecosystems, species, wild populations, local varieties and breeds of domesticated plants and animals are shrinking, deteriorating or vanishing. The essential, interconnected web of life on Earth is getting smaller and increasingly frayed."[15]

The IPBES explicitly calls out the culprits for what can only be considered an unprecedented ecocide.[16] Based on a thorough investigation of available evidence, the authors conclude that the five direct drivers of the biodiversity crisis, in ranked order, are: (1) changes in land and sea use; (2) direct exploitation of organisms; (3) climate change; (4) pollution; and (5) invasive alien species. In other words, agriculture and fishing, key contributors to changes in land and sea, are the primary causes of the deterioration of nature. The IPBES study documents the inexorable destruction of biodiversity as capitalist modes of agriculture have spread a small range of cash crops and livestock across the planet, consuming forests and other nature-rich ecosystems.

The panel notes that food production has increased dramatically since 1970, with 100 million hectares of land having been put into agricultural production between 1980 and 2000, most of which has taken place in the tropical countries of Latin America (in the form of cattle ranching) and in Southeast Asia (mostly in palm oil plantations). Roughly half of this expansion has taken place through the destruction of intact forests. Despite this massive growth, over 10 percent of people alive today remain undernourished. The report also notes that 23 percent of land areas have seen a reduction in productivity due to land degradation. In addition, since 75 percent of global food crops rely on animal pollination, hundreds of billions of dollars of crop output are now at risk due to pollinator loss as insect numbers plummet around the planet.

Building the resilience of food systems and of the natural world upon which we all depend will require transforming the dominant organization of agriculture, which is brutalizing both people and planet. It will require fighting a global capitalist system based on constant expansion through feckless exploitation of fossil fuels. And it will require fighting food imperialism. Roughly 50 percent of carbon emissions are produced by industrial agriculture, if one includes all aspects of capitalist food systems, from deforestation to transportation, processing, and waste.[17] Transformation of capitalist food systems is consequently a key element in what must be a much broader program of ecological, economic, and social reconstruction.

For Green New Deal planners in wealthy nations, the pandemic catalyzed calls for a People's Bail Out of many sectors of society, and yet the thoroughly destructive human and environmental impact of capitalist agriculture on a global scale went relatively underacknowledged.[18] The failure to center agriculture and the fate of farmers—and most importantly the parlous situation of the millions of peasants in the Global South who produce the vast majority of humanity's food on small plots of land—is testament to the flawed character of many plans for ecological reconstruction. Indeed, unless discussions of the Green New Deal center the survival of the world's farmers, they will be nothing but another neo-colonial assault on the Global South. Decolonizing agriculture must be a fundamental element in any environmentally and politically viable plan for global ecological reconstruction.

Small farmers face a constant onslaught today. According to the agronomist Eric Holt-Giménez, more than 70 percent of the world's food is grown by small family farmers on less than 25 percent of the world's arable land.[19] The 30 percent of the world's food not produced by small-scale farmers is grown on huge farms run by massive agribusinesses. Four companies—Archer Daniels Midland, Bunge, Cargill, and Louis Dreyfus—control 90 percent of the global grain trade.[20] These monopolistic enterprises are consolidating vertically and horizontally to control nearly every node in the global food system, from seeds to fertilizers, processing, packing, distribution, and retail. To call these corporations Big Ag is to engage in significant understatement since they are quite literally trying to control every aspect of food.

I term this Big Ag system *agrocapitalism*, by which I mean a disparate set of technologies developed and yoked together by a scientific order directed toward capitalist exploitation in agricultural production. An iron law governs agrocapitalism's transformation of the world food system: expand ceaselessly or die. Agrocapitalism tries to gobble up the land and livelihoods of small farmers constantly. Such rural pillage, which the radical geographer David Harvey calls "accumulation by dispossession," has in fact been a primary facet of contemporary capitalism and empire—contrary to Marx's description of capitalism's sacking of the countryside as "primitive accumulation," a set of brutal acts of enclosure that established capitalism but then largely faded as the system matured.[21] Not only is this ongoing process of dispossession unjust: it also makes global food supplies far less secure. As George Monbiot points out, the integration of farming the world over into agrocapitalism "relies on ripping down circuit breakers, back-up systems, and modularity," streamlining the global food system to increase profits but also rendering the entire system more vulnerable to shocks and catastrophic crises.[22]

Land has become so cheap in much of the Global South that it can be acquired for a pittance by multinational agribusiness corporations.[23] The resulting land grabs have fueled mass dispossession of the global peasantry, millions of whom have been displaced either internally or internationally as it has become impossible to subsist in the countryside. As we will see, this situation results from the decades-long brutalization of formerly colonized countries by an international trade regime controlled by the rich countries.

Yet, while the world's population may have passed the watershed mark of being 50 percent urban, in much of the Global South the majority of people still live in rural areas, and even urban dwellers often migrate backward and forward to rural hinterlands. These realities should dictate a totally different approach to planning for ecological reconstruction in the Global South. As the radical South African think tank Alternative Information and Development Centre (AIDC) puts it in its plan for a million climate jobs, "If you read material produced by our sister climate jobs campaigns in the global North, in Britain, Norway, Canada or France, you will see that they start with renewable energy, transportation and building conversions. In South Africa, we are going to start with farming and people in rural areas. At the most fundamental level, climate change is going to affect farming, and the growth of enough food to meet the needs of all, the most."[24] Focusing first and foremost on the climate crisis as a crisis of rural livelihoods and agricultural sustainability is, in other words, a key element in plans for ecological reconstruction.

Rural communities in the Global South thus feel climate chaos differently than the rest of the world, but the differences don't end there, as there are also disparate experiences *among* and *within* rural communities. As we have already seen, for example, rural women play a vital role in building rural livelihoods, yet tend to bear the brunt of poor working conditions, food insecurity, and climate-related stresses like drought and floods. As the AIDC puts it, "Communities that are already struggling to survive off the land—either through subsistence farming, or through paid labor, are going to feel the impact of climate change very starkly." Planning for ecological reconstruction, then, must be centered on questions of sustaining rural livelihoods and fighting gender inequality.

Environmental transformation in the Global South in general and in rural areas in particular must come from below, from the people of the countryside and the social movements they form and sustain. The AIDC's plan for transformation in South Africa rightly notes that "we cannot make governments act on climate change without a conscious, organized and determined mass movement from below." This implies that the specific demands for ecological reconstruction grounded in the rural zones of the Global South must be determined by people who live there themselves. That said, activists with the AIDC have no hesitation about challenging the hollow and insulting remedies that international aid agencies have peddled to people faced with climate chaos: "It is not enough simply to say that rural people need advice on 'adaptation,' and that we must increase 'resilience.' Climate change will destroy farms all over the world." The first section of this chapter explores the ways in which these precise international agencies, and the global capitalist system as a whole, have helped to undermine the agricultural and social resilience that farmers in the Global South are now being enjoined to shore up.

Contrary to the long-standing prejudices about rural backwardness and lack of organizational capacity that determine dominant policy prescriptions, rural people and farmers around the world have organized some of the most powerful and global political organizations that currently exist. The assault on the global countryside has been resisted heroically by movements such as the transnational peasant confederation La Via Campesina. The second half of this chapter profiles the strategies through which the world's peasantry has fought back against the obliteration of rural life detailed in my discussion of rural dispossession. Key to this struggle are the efforts of small-scale farmers to challenge the dominant ideologies of capitalist agriculture, including the idea that big farms are more productive than small ones, that technological solutions such as synthetic fertilizers and genetically modified organisms are the answer to the challenges facing agriculture today, and that the free market will ensure the most efficient use of resources in agriculture and other areas of human endeavor.[25] Against these dogmas, farmers and farmworkers across the Global South are experimenting with the set of farming practices known as *agroecology*, developing a science of sustainable agroecosystems and the social movements to support and spread this grassroots science.

Industrial agriculture destroys the land. Karl Marx recognized this long ago, as did the pioneers of the organic movement in the early twentieth century, but technological optimism and capitalist triumphalism ensured that this became a dissident perspective in wealthy nations by the end of the twentieth century. And today, our soil is dying. Despite claims that the dominant capitalist agricultural system produces food more efficiently and in higher quantities than other systems, about 33 percent of arable land has been destroyed since 1975, as erosion of plowed fields consumes ten to one hundred times the amount of soil being formed globally.[26] Today, 24 billion metric tons of fertile soils are lost from farmland around the world each year.[27] As geologist David Montgomery reminds us, the decline and fall of human civilizations across the eons can be linked to their degradation of soils—the frequently unacknowledged but nonetheless foundational terrain for their prosperity.[28] Notwithstanding our tendency today to imagine that technological innovation will provide solutions for every problem we confront, there is no getting around the problem of consuming soil, arguably our most valuable resource and the one on which all life depends, faster than we generate it.

A change is gonna come, one way or another. The status quo in the global food system is untenable: according to the Food and Agriculture Organization, the world has only about sixty harvests left given the current rate of soil depletion.[29] As climate change–linked intense droughts and extreme rainfall make farming increasingly challenging in many parts of the world, many more rural people will face slow starvation and dispossession. Life in the countryside is becoming increasingly difficult to sustain. The rural movements that constitute one of the most important expressions of environmentalism from below are fighting to reverse rural dispossession and institute a decolonized agriculture that will play a pivotal role in confronting the compound environmental and social crises of the countryside today. The solution to the conjoined climate crisis and global food crisis will never be a purely technological fix, but must instead involve revolutionary social transformation. Land must be taken away from local elites and from transnational corporate agriculture and restored to the people. Peasant communities must be able to decide what they will grow and how they will grow it. And the exploitative system of free trade established during the neoliberal era must be rolled back so that national food production in Global South countries can be protected from powerful, subsidized industrial agricultural businesses in the West. In the words of militants in La Via Campesina, it is time to globalize struggle, and to globalize hope!

## The Fourth Industrial Revolution in Agriculture

Here come the robots. While a 2018 accident in which a self-driving Uber car hit and killed a pedestrian in Arizona ensured that autonomous automobiles will not appear on roadways in the US for the time being, robots are taking over agriculture in a surprising futurist twist.[30] Challenging our tendencies to see rural areas as backward in temporal and cultural terms, robo-tractors may soon be plowing the nation's fields and harvesting its crops.[31] Bear Flag Robotics, a California-based firm, is developing the actuators and sensors that will allow existing tractors to guide themselves through fields, stepping up already-available auto-steer systems that use GPS receivers to keep rows of crops straight but still require a person at the helm.[32]

From robo-tractors to genetically engineered seeds, technological solutions are touted by proponents of agrocapitalism as the solution to the food crisis I outlined above. Make farming more efficient through automation, the line goes, and we will be able to produce more food that will feed more hungry people. Yet the technological transformation of farming is just one part of capitalism's broader tendency to increase the automation of labor. This shift in the ratio of automated labor to living labor has been an intrinsic feature of industrial capitalism from its inception—think about the replacement of handloom weaving by machines in nineteenth-century Britain. Karl Marx in fact described the factories of his time as a kind of "vast automaton" driven inexorably forward by a "self-acting prime mover": capital.[33]

From robotic assembly lines to GIS systems in mining, in recent decades there has been a staggering acceleration of this tendency toward automation, and a concomitant replacement of human laborers with machines.[34] The result has been a massive growth in the number of people who are no longer needed in production, a group often called the "global reserve army of labor." This trend can put power in the hands of employers, since there is always a precariously employed person willing to work for less wages. But technology is not necessarily a perfect fix for capital's problems. Indeed, Marx famously argued that automation tends to produce a falling rate of profit; this is because capitalists must constantly invest in new machinery to out-compete one another, eroding the profits gained by individual capitalists as they adopt more efficient technology. His analysis seems to have been borne out in recent decades by low rates of growth.[35] As we will see, the technological transformation of farming so often celebrated today as a solution to the food crisis is in fact intensifying the underlying economic, social, and environmental contradictions of agrocapitalism.

GPS has been in use on the farm for over two decades, but behind the drive to deploy high tech in agriculture today, according to industry spokespeople like Bear Flag CEO Igino Cafiero, is a crushing shortage of labor. There just aren't enough farmworkers to harvest all the crops in California, Cafiero claims. Across the country, in Florida, Gary Wishnatzki, a strawberry farmer and founder of the robotic berry-picking firm Harvest CROO, makes a similar argument. Unlike staple crops like wheat, corn, and oats that are harvested by mechanized means, such as the tractors and combines introduced early in the twentieth century, berries are still picked by hand. Workers spend hours bent over in the fields, delicately culling only the ripe berries and leaving green ones to mature on the vine. As a result of demographic changes, including an aging and more affluent population in Mexico, home to the majority of the US's migrant farm labor, the US lost over a million migrants between 2007 and 2014. Despite the pervasive xenophobic rhetoric in the US today, Wishnatzki says that there are far fewer potential berry pickers available today than in previous years: his firm's Berry 5 robopicker is, he argues, about food security.

Agriculture is a prime field for the experimental technologies driving what proponents call the "Fourth Industrial Revolution," or 4IR. For key boosters such as the World Economic Forum (WEF) and the McKinsey Global Institute, 4IR involves technologies like artificial intelligence (AI), machine learning, robotics, sensors, cloud computing, nanotechnology, 3D printing, and the internet of things.[36] As the WEF puts it in breathless terms, 4IR will see "access to technology spread like wildfire. Almost anyone will be able to invent new products and services cheaply and quickly. The business models of each and every industry will be transformed."[37]

While, as we shall see, there is good cause to question WEF's booster-ism of 4IR, its observation that 4IR will upend existing industries certain-ly seems accurate in regard to agriculture. For example, Harvest CROO's Berry 5 robopicker and similar experimental devices employ a bevy of converging technologies, including GPS, artificial intelligence, machine vision, robotics, and Big Data.[38] Linked to the burgeoning use of these technologies in agriculture are extreme forms of genetic engineering such as CRISPR, which allows scientists to "edit" a plant's genome in order to produce desired traits such as drought resistance without transferring genes from one organism to another, as in the current method of creating genetically modified organisms.[39]

Animating these disparate developments is a recognition that envi-ronmental disruption is already transforming the lives of farmers, gener-ating unpredictable weather patterns, shorter growing seasons, extreme temperatures, droughts, and increased problems with pests and crop dis-eases. *Climate-smart agriculture*, a baggy term that includes many 4IR technologies, is touted as a solution to climate chaos. Exemplary of this discourse, a new initiative of the Gates Foundation called the Bill and Melinda Gates Agricultural Innovations LLC, or "Gates Ag One," prom-ises to "bring scientific breakthroughs to smallholder farmers whose yields are threatened by the effects of climate change."[40] Like makers of agricul-tural robots such as Harvest CROO, Gates Ag One stresses that the issue of food security is fundamental.

But lurking in the subtext of such expressions of concern about food supplies are undeclared fears about political insurrection. Food riots, a frequent occurrence in the early years of the Industrial Revolution, have become increasingly common in recent decades. In 2008, for example, Egyptian families had to queue up for hours each day to get bread rations as the country coped with a serious grain shortage. For poor Egyptians such as Asma Rushdi, who had to wait to buy her small ration of sub-sidized bread from a state-owned bakery, "bread is everything."[41] That year, tensions over access to bread led to riots. A wave of popular anger over dramatic price hikes for basic foodstuffs such as rice, cereals, cooking oil, and sugar also helped spur the uprisings that became known as the Arab Spring.[42] As radical political ecologist Andreas Malm put it in his discussion of the Egyptian revolution of 2011, "inability to access food has a famous capacity to radicalize. . . . The ruling regime is perceived less as a guarantee than as a threat to the bodily metabolism of its people; it therefore risks losing all legitimacy, while the masses might feel that they have nothing to lose."[43]

What do these technological innovations mean for the world's farmers, and for workers more generally? Even tech boosters like the WEF note that 4IR could intensify existing forms of capitalist dispossession: "Disruptive technologies could help distribute food, wealth and data, reduce hunger and waste, and empower farmers to produce more valuable, climate-resilient and nutritious foods for their clients. Or they could spur a consolidation of the food sector, allowing a few companies to dominate the market, limiting food choices and expanding bad practices rather than correcting them."[44]

These concerns about monopolistic tendencies in today's food system are grounded in structural dynamics of agrocapitalism. Capitalist bosses have a long history of employing fossil-fueled automation technologies to put workers out of jobs, in the process generating mass unemployment and immiseration, as well as a declining rate of profit. Even the WEF's whizzbang video on 4IR ends with the vexing question, "How do we avoid a world of joblessness, low productivity, and inequality?"

While it may be quite satisfying to hear an über-capitalist organization like the WEF worrying about the destructive potential of automation, it is important to remember that the social impact of even the most powerful technologies is not predetermined. As Aaron Bastani puts it in *Fully Automated Luxury Communism*, "How technology is created and used, and to whose advantage, depends on the political, ethical and social contexts from which it emerges."[45] In his appealingly titled book, Bastani imagines a utopia of collective well-being and unbounded leisure time snatched from the jaws of 4IR developments, from an AI takeover of all labor to bespoke gene-editing therapies to cure all diseases. His point is that the impact of 4IR technologies will be determined by political struggle rather than by the technologies themselves.

Some commentators have argued that these novel agricultural technologies ought to be appropriated and transformed by rural social movements.

For instance, the Out of the Woods collective suggests that revolutionary peasant organizations might imagine "a more capacious set of knowledge practices" that would "adapt techniques from the formal sciences through bricolage."[46] They ask, for example, what would happen if we treated transgenic crops as part of the genetic commons rather than enclosed, proprietary forms.

This idea of appropriating and decommodifying transgenic crops sounds appealing. Yet revolutionary peasant groups must not forget that agricultural technologies such as GMOs are expressly designed for domination. They are interlocking elements of what I have been calling *agrocapitalism*. Agrarian social movements cannot simply hijack pieces of this system for their own use since these pieces carry exploitative violence within their very design. The science that produces these technologies thrives on monopolistic, value-generating knowledge forms that perpetuate political and social hierarchy. The new 4IR technologies being developed for agriculture are of a piece with the long tradition of extraction that characterizes imperial science.[47]

Even the most cursory analysis of the impact of 4IR-supporting entities like the Gates Foundation—with its bankrolling of agricultural data-mining organizations like Digital Green, advanced gene-editing technologies like CRISPR, and Stress Tolerant Rice varieties for Africa and South Asia—underlines the malign impact of 4IR technology on the global peasantry today.[48] We should be extremely wary of the tendency to imagine that these technologies will be easily borrowed by rural revolutionaries. Rather than backing agrarian social movements' adoption of such capital-intensive technologies, we need to support these movements' efforts to develop radical alternatives.

According to political theorist Robert Biel, the extractivist orientation toward the natural world that characterizes agrocapitalism is grounded in alienation, which "in its narrower economic sense means separating us from the conditions and product of our labor and, in a wider sense, a psychology which cuts us off from nature."[49] For Biel, the dominant Western orientation to nature also cuts us off from the consequences of our acts, a fact abundantly evident in the disconnect that prevents the leaders of Western nations from addressing climate breakdown seriously even as the world's coral reefs bleach, forests burn, seas rise, and carbon levels spiral. Extractivism, which Biel suggests might equally accurately be termed *exterminism*, led to the invention of agricultural approaches organized around artificially simplified and homogenized technologies and systems. The complexity of traditional forms of agriculture the world over, which were predicated on working with the emergent organization of the natural world, was thus abandoned for a system of industrial agriculture that was based on monoculture, has high amounts of external inputs, and is dependent on fossil fuels. Karl Marx predicted the outcome of this approach, writing that "all progress in increasing the fertility of the soil for a given time is a progress towards ruining the more long-lasting sources of that fertility."[50] Today, with commentators talking about "peak soil" as topsoil vanishes at up to 50 tons per hectare per year, one hundred times faster than its formation rate, the accuracy of Marx's damning assessment of the underlying dynamic of capitalist agriculture is clear.[51]

## Imperialism and Food

To understand why contemporary agrocapitalism is so destructive of soils and of planetary ecosystems more broadly, we need to trace the role of food systems in circuits of capitalist exploitation and empire over the last century. After World War Two, the United States retrofitted wartime factories to produce tractors rather than tanks, pesticides and fertilizers rather than chemical weapons. Industrial agriculture was born, and food production soared to unprecedented heights. These policies were a response to the deep crisis of the Great Depression, when prices for staple foods crashed. Faced with slumping demand, farmers tried to grow their way out of poverty by producing as much food as possible. Food flooded constricted markets, leading to even greater price drops for agricultural commodities. Postwar government policies in the US such as production quotas and price supports for farmers were intended to address this crisis, yet, as the agroecologist Eric Holt-Giménez argues, they simultaneously continued to support surplus food production.[52] In the initial postwar years, the US sent much of this surplus food to the war-devastated countries of Europe, but as these countries rebuilt and restored their agricultural productivity, they too began to overproduce food. Chronic surpluses quickly built up. What to do with all this food?

Federal authorities soon arrived at a solution. They would send surplus food—labeled as "food aid"—to the countries of the Global South, many of which had only recently achieved independence after centuries of colonial domination. Governments there sold this food at low prices in the local currency, generating revenue that could in turn be used for public works.[53] Such "aid"—codified as Public Law 480 in 1954—was a powerful propaganda tool in the Cold War with the Soviet Union, helping to support the image of the US as a generous benefactor and defender of democracy around the world. Yet food aid was driven not by philanthropy, but rather by a determination to cement the allegiance of governing elites in Global South countries. As George McGovern, then director of the Food for Peace program, put it, food aid "is a far better weapon than a bomber in our competition with the Communists for influence in the developing world."[54]

Food aid came at a pivotal moment for Global South nations. After centuries of imperial extraction often leading to famines that killed millions, many countries that freed themselves from colonization had placed priority on the development of their own agricultural sectors.[55] One of the primary means of doing so was through land redistribution, which took fertile terrain out of the hands of the powerful landlords, who had often functioned as de facto tax collectors for imperial overlords, and gave it back to the peasants who actually worked the land. From China to Burkina Faso to Cuba, revolutionary campaigns for land redistribution swept across the world in the period from 1945 to 1970. But food aid often drove peasants in Global South countries out of business, since they could not compete with the low prices charged for government-subsidized food exported from the US. Local food production was undermined around the world, with newly independent nations, in the space of a few decades, going from supplying food to the Global North to being dependent on rich nations for food. This situation helped legitimate discourses of "development," thinly veiled racist ideas about the innately backward character of former colonized nations that helped prop up US empire during the "American century."

Among the major modes of so-called development was the Green Revolution, a campaign to spread capital- and energy-intensive agriculture in the countries of the Global South. Alarmed by the wave of land redistribution and anti-colonial uprisings, theorists of development such as the prominent American economist Walt Rostow saw the Green Revolution as a key policy tool. The goal for Rostow and other architects of US power was to move "traditional" societies toward "modernization." This meant that the nations of the Global South were to abandon long-established farming methods for what were seen as increasingly productive technologies that would allow farmers to grow more food, generating higher profits and economic "takeoff" that would stamp out the appeal of socialism in the Third World. The philanthropically framed goal of this campaign to disseminate capitalist agriculture was to achieve food security in the "developing" countries. The Green Revolution was thus depicted as saving the world from hunger by ramping up agricultural production—yet, contrary to propaganda about the cornucopia of food production it generated, the Green Revolution actually produced just as many hungry people as it saved.[56]

On a technological level, the Green Revolution was based on the use of dwarf varieties of Mexican wheat developed by the US agronomist Norman Borlaug. These newly engineered varieties of wheat, along with similar strains of rice, corn, and other staple crops, were characterized by extremely high yields. Farmers growing them were able to harvest bumper crops. They seemed like miracle crops—what was not to like? For one thing, the Green Revolution had economic and ideological backing from the Rockefeller and Ford Foundations, organizations grounded in, respectively, petroleum and automobile manufacturing. The backing of these foundations was no coincidence, for the high-yielding crops that the Green Revolution spread around the world only worked when combined with heavy applications of fossil fuel–based fertilizers and pesticides. The campaign thus helped spread consumption of petroleum-based technologies to countries that had not been part of the global circuits of fossil capitalism.

The Green Revolution also radically diminished the biodiversity that had characterized peasant agriculture. In the Punjab region in northwest

India, one of the centers of the Green Revolution, government policies that were rolled out beginning in the 1960s encouraged farmers to abandon their traditional heterogeneous mix of crops to grow only the high-yield rice, wheat, and cotton developed by the powerful Western interests. The new "miracle" seeds could produce far bigger yields than Indian farmers had ever seen. But there was a catch: the crops also required far more water. Farmers could no longer rely on natural rainfall but had to dig wells and irrigate crops with groundwater. Today, farmers like Sundeep Singh in the Punjabi village of Chotia Khurd must dig down hundreds of feet to reach the water in a dramatically depleting water table.[57] The equipment that he and his neighbors use to pump up water for irrigation costs thousands of dollars. Already deeply in debt, Singh and farmers like him are no longer able to take out loans from banks and must rely on "unofficial" lenders like local businessmen, who charge double the rate of interest charged by banks. Sundeep Singh's situation is emblematic of the global plight of farmers who embraced the Green Revolution, only to find themselves caught in a spiraling debt and ecological trap.

This trap involves corporate control of the most basic elements of agriculture: seeds. Since the hybrid seeds of the Green Revolution did not reproduce true to type, farmers could not save their seeds each year for subsequent replanting. For seed-supplying companies, hybrid varieties were a proprietary technology anathema to the seed-saving and -sharing traditions of most of the world's farmers. The seeds of the Green Revolution consequently made individual farmers—and entire nations—dependent on the seed-supplying corporations. It was a frontal attack on the autonomy of the world's peasantry. Moreover, the combination of hybrid seeds, hefty fossil fuel–based inputs, and large-scale machinery was not just a form of enclosure of the seed commons: it was also expensive, so only the most well-off farmers in nations such as India, Pakistan, and Mexico could afford to take advantage of Green Revolution crops. This meant that richer farmers acquired competitive advantage, pushing their poorer neighbors into debt and destitution.

The result—combined with the policies of food "aid" discussed above—was a massive immiseration and displacement of the global peasantry, who migrated to fragile terrain on hillsides or the forest frontier to grow subsistence crops—or gave up farming entirely to participate in a global exodus to cities. The world's rural population decreased by 25 percent from 1950 to 1997. Today, one out of every eight people alive—a total of one billion human beings—lives in urban slums.[58] The 63 percent of the world's urban population that now lives in or on the margins of the sprawling megacities of the Global South suffers from chronic hunger and malnutrition, even while the global food system churns out unprecedented quantities of food—nearly 50 percent of which is wasted.

The Green Revolution's global diffusion of high-yielding hybrid crops also displaced thousands of local varieties of maize, rice, and wheat, leading to a 90 percent loss of agricultural biodiversity.[59] High-yield seed varieties bred to "tolerate" large doses of herbicides that killed off competing plants helped spread new toxins around the world. Instead of using the mixture of plant varieties known as *intercropping*, a technique that allows plants to borrow pest immunities from their neighboring plants, the agricultural technologies of the Green Revolution embraced what was effectively a program of chemical carpet-bombing. The push to export these new high-yield varieties of wheat to postcolonial nations, and the political relations of dependency woven into their technical matrix, was led by a new global network of agricultural research centers now known as the Consultative Group on International Agricultural Research, an initiative developed during the postwar years but formally consolidated in 1971 with support from the World Bank as well as the Rockefeller Foundation and UN organizations like the Food and Agriculture Organization (FAO). These research centers dispatched a host of well-paid consultants around the world to encourage farmers to adopt capital- and energy-intensive industrial agriculture.

The sting in the tail of these efforts at "modernization" became evident in the early 1970s, when a major food crisis affected countries across Asia, Africa, and Latin America. The immediate cause of this crisis was an intense drought that impacted entire regions of the Global South, but the impact of this disturbance in natural cycles was intensified by spiraling prices for fossil fuels, which produced a global shortage of the fertilizers whose adoption the Green Revolution had proselytized. With US grain stockpiles depleted as a result of Cold War food aid policies, severe food shortages arose and food prices soared across much of the Global South.[60] The dominant policy response to the food crisis was to double down on the Green Revolution's emphasis on increasing agricultural production. US secretary of agriculture Earl Butz famously called on farmers to plant "fencerow to fencerow," and simultaneously supported the consolidation of farms into giant corporate enterprises by arguing that farmers should "get big or get out."[61] Rather than concluding that the trend toward monoculture and monopoly that characterized the Green Revolution was increasing the fragility of global food systems, elites decided that the solution was an intensification of agrocapitalism.

But instead of resolving the contradictions of the global system of agrocapitalism, these policies intensified the food crisis. The two decades before the turn of the century saw the lives of farmers across the Global North and South grow increasingly precarious as commodity prices crashed in the wake of ramped-up production. Entire rural communities were hollowed out in the US Midwest, as farmers who had borrowed heavily to expand their production in the 1970s struggled to cope with plunging prices for their crops and a downturn in exports. The situation was so bad in rural areas of the world's imperial hegemon that Willie Nelson and other musicians organized the first Farm Aid concert in 1985 to raise relief money for struggling family farmers.[62] Meanwhile, across the Global South, small-holding farmers were confronted with a vicious wave of debt imperialism. In the 1970s, banks in the US and European Union were awash with capital generated by the decade's spikes in oil prices. Looking for good returns on this money, they lent enormous sums to postcolonial nations anxious to jump onto the development bandwagon. Global South nations soon found themselves shouldering unsustainable debt.

Mexico, the nation that had initially hosted experiments in Green Revolution seed technology, was ground zero for this debt crisis. During the 1970s, Mexican governing elites had borrowed tremendous sums from international lending institutions like the World Bank, as well as private banks in the US, Europe, and Japan. Mexican borrowing increased sixfold during the years between 1973 and 1981.[63] These loans came to a sharp and shocking end, however, when the US Federal Reserve engineered what has come to be known as the "monetarist counter-revolution."[64] In response to high inflation rates in the US, the Fed jacked up interest rates precipitously, refunneling global financial flows into the dollar and depriving nations like Mexico of the cheap loans on which they'd become dependent. On August 20, 1982, the Mexican government, having paid back considerable amounts of these loans in the first seven months of the year, stated that it could not pay any more and declared a six-month moratorium on loan repayments. By the end of August, the head of the International Monetary Fund told Mexican authorities that the US Treasury, the Bank of International Settlements, and the Bank of England would bail Mexico and other debtor nations in the Global South out. But these loans were not philanthropy: they came with stringent conditions that included the dismantling of the grain reserves that Global South nations had built up after independence, a halt to their efforts to grow basic food crops, and a new agricultural orientation based in specialization in "non-traditional" export crops—fancy food and flowers for gourmet markets in the rich countries. As state marketing boards were replaced by private buyers, peasant producers found themselves at the mercy of speculative sharks.

Across the world, tens of millions lost their land under the impact of trade liberalization and export agriculture.[65] Without access to land, these people had no recourse other than to sell their labor in countries whose industrial sectors tended to be anemic and hyper-exploitative. Cast onto the morality-free free market, they found that they often could not afford to buy food. "Free markets" were thus inextricably linked to food riots in the sprawling slums that the capitalist food system helped produce around the world. As food analyst and activist Raj Patel argues, food riots exploded in the 1980s, the 1990s, and again during the Great Recession of 2007–8 not because the world was short of food but rather because food was unevenly available.[66] Between 1976 and 1992 there were 146 protests against IMF-sponsored programs in thirty-nine different countries across the world.[67] This global wave of revolt against austerity came to be known as the "IMF riots." The most famous of these riots took place in Venezuela—the Caracazo of 1989—kicking off a social upheaval that eventually brought Hugo Chavez to power and that created the Bolivarian Revolution.[68] Such uprisings index not simply anger at rising food prices but also popular rage at the failure of the state to provide the most basic elements necessary for people to survive. Food riots are hence an expression of moral outrage over the unjust system of neoliberal agriculture that, in Patel's words, sees some stuffed while others starve.

In the face of these intensifying crises, global elites doubled down on the very capitalist agricultural system whose contradictions generated the crisis in the first place. The decades leading up to the end of the century saw a push for greater agricultural trade "liberalization" through the Uruguay Round of the General Agreement on Tariffs and Trade (GATT), negotiations that culminated in the creation of the World Trade Organization. The Agreement on Agriculture, adopted upon the official establishment of the WTO in 1995, aimed to open up agricultural markets by reducing tariffs imposed by countries around the world on agricultural goods. The justifying ideology for this move was "food security": if there were no tariffs jacking up the price of food, the reasoning went, people around the world would be able to buy it at the cheapest possible prices. The magic hand of the market would ensure that cheap food would flow to hungry people. All hail the magic hand. But WTO agreements allowed wealthy countries to continue subsidizing agricultural production—to the tune of $300 billion in industrialized countries—while simultaneously rolling back the tariffs that protected farmers in the Global South. For agrarian social movements and their allies in the Global Justice Movement, the establishment of the WTO was nothing but a new round of imperialism in which the rhetoric of "food security" obscured mass dispossession for the global peasantry.[69]

In 1996, the governments of the world met at the UN World Food Summit in Rome to address the ongoing problem of food insecurity. At this point, a country like Côte d'Ivoire, which had been nearly self-sufficient at the previous World Food Summit in 1974, was only able to meet half of domestic food demand. Similar patterns prevailed across the rest of Africa and Latin America. Although government representatives pledged to eradicate hunger in all countries and to "promote equitable access to productive and financial resources," the concrete solutions they proposed in 1996 were virtually identical to those advanced in the 1970s: increased food production and further trade liberalization.[70] In other words, although they were willing to acknowledge the dramatic increase in hunger despite the "miracle" seeds of the Green Revolution, elites were not willing to consider genuine alternatives to the agrocapitalist system that was producing hunger.

The principal difference in their response lay in the hopes invested in a new technology: genetically modified crops. The technique behind GM crops, recombinant DNA, allowed scientists to harness the process through which bacteria exchange mobile bits of genetic information among themselves. Whereas traditional breeding methods, including the hybrid seeds of the Green Revolution, were limited by the rules of sexual compatibility, the new techniques of genetic engineering enabled biologists to move genetic information across the barriers of species and genus. DNA could now be transferred from plants to animals to bacteria and back again. Genes were being turned into factories. According to the historian of life sciences Melinda Cooper, legislative and regulatory changes made in the US during the 1980s "relocated economic production at the genetic, microbial, and cellular level, so that life becomes, literally, annexed within capitalist processes of accumulation."[71]

This shift into what has been called *biocapitalism* was catalyzed by declining profit rates for US petrochemical and pharmaceutical corporations from the export of fertilizers and pesticides, partially as a result of the environmental and economic devastation wrought by these Green Revolution technologies.[72] These corporations reinvented themselves as purveyors of genetically engineered seeds, often designed to work with the very same toxic pesticides that had been in use before the biocapitalist turn. Taking advantage of new patent laws such as the Bayh-Dole Act, which allowed US universities to acquire patents on inventions resulting from government funding and for private firms to receive exclusive licenses, these industries reacted by shifting investment into new genetic technologies.[73] Using licensing agreements with biotech start-ups or forming their own in-house research units, US corporations like Monsanto reinvented themselves as purveyors of new, "clean" life science technologies. The WTO added agricultural provisions and "trade-related aspects of intellectual property rights," or TRIPS, giving a handful of mega-corporations like Monsanto and Bayer the right to disseminate transgenic crops across the planet while banning Global South nations from creating their own copies of GMOs.

As Cooper puts it, rather than mass producing chemical fertilizers and herbicides, agrocapitalism "displaced its claims to invention onto the actual generation of the plant, transforming biological production into a means for creating surplus value."[74] The media hype that attended this shift was deafening: at the heights of the tech euphoria of the 1990s, the biotech industry was celebrated as holding the key to ending hunger, solving the biodiversity crisis, and overcoming the endemic problem of industrial pollution.

But toxic pollution, hunger, and the extinction crisis only got worse in the subsequent decades. A major reason for this is path dependency, or the process through which an established paradigm acquires an inertia, so that past choices shape future ones. One good example of such path dependency, according to Robert Biel, is the feedback loop of the Green Revolution, where chemical-intensive agriculture undermines the health of soils and ecosystems more broadly, which leads to declining yields, leading farmers to apply more chemicals, further damaging soils and necessitating greater applications of toxic chemicals.[75] But path dependency also applies to the close interdependence between chemicals and seeds. As we have seen, this arrangement was enormously profitable for agrocapitalism and also played a powerful role in US food imperialism. This chemical-seed combination was not simply cast aside with the advent of agricultural biotech, but rather re-engineered as agrocapitalist corporations developed new plant varieties to go with already-existing pesticides.

A clear example of this is the story of glyphosate, a carcinogenic herbicide developed by the Monsanto corporation during the 1970s that has become the world's most widely used weed killer, commonly used on farms and in gardens under the folksy-sounding name "Roundup."[76] Glyphosate may be a great weed killer, but it is also great at killing crops in general. So in 1996 Monsanto introduced the Roundup Ready soybean, a genetically engineered crop resistant to glyphosate. In the years since, Roundup Ready cotton, maize, and other crops have made their debut. Advocates of this crop-herbicide combination argue that it is environmentally beneficial since it allows farmers to replace more harmful herbicides with smaller concentrations of glyphosate, and also because it permits farmers to avoid tilling their fields in order to kill weeds.[77] But in the years since its introduction it has become clear that the exclusive application of Roundup to millions of acres of monocultural cropland is essentially an experiment in breeding glyphosate-resistant superweeds. Farmers in the US, where the majority of these superweeds have developed, are now back to tilling and spraying more toxic herbicides on their fields. The Roundup story is a tragic exemplar of the way in which the alienated approach to nature that characterizes agrocapitalism backfires, as a reductionist view based on mastering a fragmented natural world generates precisely the opposite results as those advertised. Escaping this toxic vortex will involve breaking the inertia built into agrocapitalism by adopting alternative approaches to nature and farming.

## Fighting Food Imperialism

On September 10, 2003, a South Korean farmer and peasant organizer named Lee Kyung Hae climbed atop a truck near the barbed wire surrounding the WTO Ministerial Meeting in Cancún, Mexico, flipped open a small pocketknife, and stabbed himself in the heart. He died two days later. In a pamphlet published earlier that year, Lee had written:

> My warning goes out to all citizens that human beings are in an endangered situation. That uncontrolled multinational corporations and a small number of big WTO Members are leading an undesirable globalization that is inhumane, environmentally degrading, farmer-killing, and undemocratic. It should be stopped immediately. Otherwise the false logic of neoliberalism will wipe out the diversity of global agriculture and be disastrous to all human beings.[78]

Tragic as Lee's death was, it nonetheless became a rallying call for La Via Campesina (LVC), the transnational peasants' organization to which he belonged, and other rural social movements around the world. Within days of his death, the chant "We are Lee!" had been translated into dozens of different languages, as tens of thousands of peasants marched in countries around the world in mourning and solidarity with Lee and to demand sustainable agricultural policies.[79] Lee's suicide thus became a sign of the crisis faced by peasants, Indigenous people, and landless workers—as well as a symbol of their radical challenge to the global status quo.

Prior to the creation of LVC, peasants and other rural people were entirely absent from the international stage where agricultural policy was increasingly being determined, according to Nico Verhagen of the LVC Secretariat.[80] NGOs claimed to represent the world's peasantry in emerging global forums like GATT and the WTO, but peasants themselves remained essentially a silenced subaltern. But in 1993, agrarian social movements from around the world converged at a meeting in Belgium. In *Globalize Hope*, a documentary about the history of LVC, Honduran farmworkers' union leader Rafael Alegría explains the factors that precipitated this convergence: "By 1992 we were seeing the deployment of structural adjustment and neoliberalism in agriculture. . . . We concluded that these policies were disastrous and were leading to the destruction of the world's peasantry. We decided to create an international organization of small farmers, Indigenous people, and landless people. This organization became La Via Campesina."[81] LVC brought direct political challenges to food imperialism into elite spaces. Yet for Francisca Rodríguez of the Chilean National Association of Rural and Indigenous Women, who is also featured in *Globalize Hope*, LVC was not simply a response to the new forms of domination and exploitation that emerged with the organization of the WTO in the early-to-mid-1990s. Rather, Rodríguez argues, LVC built on the five-hundred-year campaign of Indigenous people, people of African descent, peasants, and popular resistance against racial capitalism and imperialism.[82]

The dominant ideology of globalization held that the leveling of tariff barriers would allow consumers to buy food on international markets, with competition, specialization, and mass production ensuring that good food would be cheap and plentiful. Yet, as we have seen, this ideology was a cloak for a new round of food imperialism. While the UN's FAO argued for *food security* in response to increasing hunger, LVC activists such as João Pedro Stédile argued that the FAO's food security framework was founded on an illusion: a humanitarian arrangement whereby states would buy and distribute food to the hungry. Yet many nation-states had by the 1990s become captives of or willing accomplices to trade liberalization treaties, and were visibly not caring for their populations. As a result, the FAO's push for food security was not solving the problem of hunger. Stédile and other activists associated with LVC argued that the hunger crisis would only be solved if people had the means to produce their own food.

It was not access to food that was the issue, in other words, but rather control of the means of agricultural production. Agrarian social movements should push governments to put in place the conditions that would allow rural people to grow their own food. LVC activists opposed the FAO's idea of food security with the concept of *food sovereignty*, the right of peoples to healthy and culturally appropriate food, produced through ecologically sound and sustainable methods. People should have the right to define their food and agricultural systems rather than having neoliberal models imposed on them by elite-controlled organizations like the World Bank.[83] The demand was essentially that international organizations and national governments dismantle the structures of food imperialism that had dispossessed increasing numbers of rural people over the last fifty years.

LVC was not simply interested in reimposing tariff barriers protecting agriculture in the Global South: the peasant activists from around the world that formed this movement of movements advanced a vision of a transformative cultural politics grounded in the democratization of sites and structures of power. LVC activists initially hoped that they could mobilize Global South nations to support radical programs such as land redistribution and related reforms such as democratic access to seed and water, counting on Global South nations to recognize that land left fallow by landowning elites was failing to contribute to national development. In countries like Brazil, an LVC stronghold, this dedication to land redistribution is enshrined in the constitution. But in recent decades LVC activists have come to realize that change will only come about as a result of autonomous mobilization by peasant organizations in a global federation. In its Marabá Declaration (2016), LVC warned of "a growing national and transnational alliance between extractive industries and agribusiness, international capital, governments, and, increasingly, even with the mainstream news media."[84] The mechanisms of governance in Global South nations have, in other words, increasingly been captured by imperial interests and powers. Political and material change can only come about through mass protest against the resulting injustices.

LVC would meet these interlinked institutions of food imperialism with networks of bottom-up solidarity. Challenging long-standing ideas about the global peasantry as outmoded and backward-looking, LVC activists from early in the organization's history established innovative new governance structures based on gender parity, equal representation of diverse regions of the world, and inclusive, horizontal practices of deliberation and decision-making.[85]

In addition to grounding itself in a radical transformation of organizing methods, LVC also set out to revolutionize agriculture itself. This was necessary, LVC activists recognized, since demands for food sovereignty made no sense if peasants did not own the seeds from which all life germinates. For Argentinian peasant leader Angel Strappazzon, the Green Revolution and the subsequent liberalization of agricultural and food markets "turned food and land into commodities subject to financial speculation."[86] Peasants had to find ways to retain ownership of seeds, land, and other elements of agricultural systems, and thereby avoid debt traps and other forms of dispossession. LVC insisted that there was a thread between the commodification of seeds and other agricultural components characterizing agrocapitalism and the Green Revolution, and what became known as "land grabs"—waves of enclosure and commercialization of land and other natural resources across the Global South during the era of the conservative counterrevolution, beginning with the debt crises of the early 1980s.[87] Agrocapitalism was dispossessing the world's rural peoples and, in tandem, annihilating biodiversity through the imposition of chemical-based monocultures and export-oriented agricultural systems that promote rampant deforestation.

As an alternative to these intertwined forms of destruction and enclosure, LVC activists came to promote *agroecology*, a phenomenon that is at once a set of farming practices, a science, and a social movement.[88] Agroecology draws on the social, biological, and agricultural sciences, integrating this cutting-edge research with the knowledge and cultures of small-scale and Indigenous farmers to generate practices that are context specific, locally adaptive, and knowledge intensive. In the terms employed by theorist James Scott, agroecology embodies the forms of practical knowledge and acquired intelligence through which rural peoples respond to their constantly changing environments.[89] Agroecology is thus the antithesis to the abstract and context-less technical knowledge that characterizes the agrocapitalist regime. Against dominant models of increasing production using extensive inputs of pesticides and fertilizers, agroecology is based on fighting for life in the broadest sense of the term. Agroecology lifts up the knowledge, culture, and embodied practices of small-scale farmers, Indigenous people, and landless workers, but it is also a fight for the lives of the worms, bees, moths, mycorrhizae, and bacteria that are essential to soil fertility. And it is also a struggle over the future of the planet, since a transition to agroecology could, according to recent assessments, lead to cuts in carbon emissions of up to 50 percent.[90]

Agrocapitalism is inherently destructive of biodiversity: by emphasizing a few cash crops in what environmental scholar-activist Vandana Shiva calls a "monoculture of the mind," agrocapitalism has drastically reduced the variety of foods we consume.[91] Today just three species of crops—rice, wheat, and maize—account for more than 50 percent of human caloric intake.[92] Ninety-one percent of maize varieties, 94 percent of peas, and 81 percent of tomatoes have been destroyed over the past century, as agrocapitalism has come to dominate the planet. According to a summary statement of the United Nations International Technical Conference for Plant Genetic Resources in Leipzig, held in 1996, 75 percent of all agrobiodiversity has been displaced because of industrial monocultures in agriculture.[93]

But industrial agriculture isn't just wiping out crop diversity: it is also stripping the land on which crops are grown. The degradation of soil stems largely from technological innovation. The chemical fertilizers and pesticides, those much-heralded promoters of cornucopian abundance in global agriculture, are the prime agents of the destruction of fertile land. We are only now coming to understand that synthetic fertilizers work like what David Montgomery calls "agricultural steroids": they prop up short-term yields at the expense of long-term fertility and soil health.[94] The much-vaunted productivity of agrocapitalism is thus based on extremely restricted metrics, which ignore not only the massive, unsustainable inputs of chemicals derived from fossil fuels but also the gradual degradation of the soil that these chemicals inexorably bring about. Agrocapitalism might best be seen as a form of slow violence against the soil, one whose short-term benefits promote an illusion of beneficence.

The agricultural science of the Green Revolution has tended to see the soil as an inert substrate, a kind of dead matrix into which chemical fertilizers can be pumped in order to make crops grow like magic. This attitude is an extension of the dominant Western, mechanistic viewpoint that sees the world as a machine and nature as dead matter.[95] Derived from Newton and Descartes, this view represents the world as composed of fixed, immutable atoms, laid out before mankind as matter to be dominated and manipulated—an attitude with a notable parallel in the European colonial attitude toward the peoples being subjugated at the time. Only now are we coming to challenge these reductive attitudes toward the material world, and to understand the fantastic complexity of the world that lies under our feet.

It turns out that soil is teeming with life. Indeed, it is perhaps the greatest locus of planetary biodiversity. For example, an average teaspoon of healthy soil has between one hundred million and one billion bacteria.[96] These bacteria decompose and immobilize nutrients, which are retained in their cells, thereby preventing soil nutrient loss. Some bacteria are also capable of fixing nitrogen, which they give to plants in exchange for carbon. Soil is replete with all sorts of organisms that engage in symbiotic relationships with plants. Earthworms and nematodes work in uncontaminated soil to break down organic matter to form humus, the carbon- and nitrogen-rich layer of living soil.

Soils rich in humus support abundant fungal life, including mycorrhizae that extract mineral nutrients from soil particles and rock fragments and make them available to plants. Mycorrhizae enter the root systems of plants, exchanging the phosphorus and other nutrients—scavenged from surrounding soil using their thin, root-like hyphae—for proteins and carbohydrates that plants generate using photosynthesis and then exude through their roots. These exudates also feed microbes that produce metabolites toxic to herbivores and pathogens, carrying out a form of natural pest control. Plants essentially cultivate and feed specialized communities of bacteria and fungi, which in turn attract predatory arthropods, nematodes, and protozoa that feast on them and then release nutrients back into the soil in a form plants can absorb. This subterranean dance of symbiotic relationships is every bit as complexly evolved as the world of flowers and pollinators that we are able to see above the ground.[97] As Montgomery points out, fifty years after the pioneer of organic agriculture Sir Albert Howard first proposed his Law of Return after observing Indian peasants' composting techniques, we finally understand that "there is a biological basis for the central role that soil organic matter plays in growing healthy crops and sustaining bountiful harvests."[98]

Agrocapitalist chemicals destroy the remarkably rich and biodiverse world harbored by healthy soil. If pesticides blight airborne pollinators like bees, butterflies, and moths, generating a global pollinator crisis, they also grievously damage the world of subterranean critters that maintain and rejuvenate soil fertility.[99] Heavy applications of chemical fertilizers can make the soil acidic and kill populations of bacteria, earthworms, and mycorrhizae by blocking the soil capillaries that supply nutrients and water to plants.[100] In addition, when plants are given all the nutrients they need through fertilizers, they cease exuding the proteins and carbohydrates on which soil microbes subsist, destroying the capacity of the soil to produce necessary nutrients and making the resulting degraded farmland increasing dependent on nitrogen fertilizers.

The practices of agroecology advocated by La Via Campesina and agrarian social movements around the world are diametrically opposed to the soil-depleting paradigm of agrocapitalism. Agroecology strives to rebuild the soil. It is based on sustaining the complex, interdependent web of life that exists on and in the soil, focusing on generating harmonious interactions between microorganisms, plants, animals, humans, and the environment.[101] While agroecology is highly diverse by virtue of being attuned to local conditions, different approaches share a few common key elements of the paradigm including the following farming practices: recycling nutrients and energy on the farm rather than introducing external chemical inputs; diversifying species and genetic resources rather than focusing on monocultures; building interactions and productivity across natural ecosystems instead of supporting individual species; integrating crops and livestock, and pest and nutrient management; and practicing the minimal and restorative approach known as *conservation tillage*, which increases soil fertility by maintaining the integrity of topsoil.[102]

The regenerative practices that characterize agroecology can be summarized in three simple but revolutionary instructions: ditch the plow, cover up, and grow diversity.[103] In place of plowing, which rips up the topsoil, conservation or minimal tillage improves soil health, reduces erosion, and decreases fuel expenditures associated with intensive mechanical tillage. Cover crops help slow erosion, improve soil health, and help smother weeds and reduce pests. Lastly, as we have seen, diverse crops can borrow one another's nutrients and pest resistance. Contrary to long-standing but blinkered claims concerning the superiority of yields produced through agrocapitalism, recent studies have shown that agroecology's principles of diversification not only conserve the soil but also ensure sustainably high yields.[104]

Agroecological farming practices can conserve and even regenerate healthy soils, the foundation of all sustainable civilizations. As Montgomery explains, the short lifespan of microbial life means that restoring soil life and fertility, thereby increasing the productivity of farmland, can happen remarkably quickly.[105] But the practice of agroecology is important not just to replenishing and maintaining the biodiversity of soils and crops. It is also key to conserving biodiversity on a much larger scale, within the besieged tropical ecosystems where the vast majority of the world's recognized biodiversity dwells.

The ecologists Ivette Perfecto, John Vandermeer, and Angus Wright argue in their important book *Nature's Matrix* that most tropical areas are highly fragmented as a result of histories of overexploitation, constituting a kind of patchwork quilt of small, intact, "pristine" ecosystems within a larger matrix of managed ecosystems—most of which are farms. Perfecto and her colleagues point out that the vast majority of conservation work focuses exclusively on the fragments of natural vegetation that remain, ignoring the matrix of primarily agricultural land within which they exist.[106] Yet for Perfecto and her colleagues, the quality of the matrix is the crucial issue in efforts to conserve biodiversity. This is because biologists have demonstrated that endangered species cannot survive on small parcels of pristine forest: to remain resilient, species must have contact with a diverse gene pool. They must be able to migrate from one wild area to the next, which means that they must be able to traverse the matrix of agricultural land within which the fragments of wilderness are embedded.

The logical conclusion is that the quality of the overall matrix is of pivotal significance to the preservation of biodiversity.

Contrary to the myopic focus of mainstream conservation efforts, creating a life-sustaining environment on farmland is key to preserving biodiversity. As we have seen, the toxic conditions on pesticide-drenched agrocapitalist farms destroys soil life. The work of Perfecto and her colleagues shows that such conditions are also inimical to the endangered species that migrate through such farmland in the world's tropical regions. Shifting away from the use of noxious external chemicals, as agroecological practices do, is thus key to maintaining biodiversity not just in the soil but on regional and even global scales. Perfecto and her colleagues in fact call for a paradigm shift in conservation, a topic which I'll discuss in greater detail in chapter 4. They propose "a reorientation of conservation away from focusing on protected areas and towards sustainability of the larger managed landscape; away from large landowners and towards small farmers; away from the romanticism of the pristine and towards the material quality of the agricultural matrix, nature's matrix."[107]

What's more, restoring the world's soils through agroecological practices could prove to be crucially life-sustaining on another level. Healthy soils are replete with remarkable quantities of carbon. The soil is essentially a giant sink for carbon. Indeed, it is the largest terrestrial carbon reservoir, second only to the oceans. This means that agroecology and similar forms of restorative agriculture can play a major role in mitigating carbon emissions. According to soil scientist Rattan Lal, worldwide adoption of conservation agriculture could put enough carbon back into soils to offset 5–15 percent of global fossil fuel emissions.[108] But Lal's estimates are on the conservative end of the spectrum: organic farming researchers at the Rodale Institute, extrapolating from results obtained on farms employing regenerative farming techniques, conclude that global adoption of these techniques could lead to sequestration of a quarter to half of all carbon emissions.[109] Montgomery explains that the difference between these estimates results from the success achieved by Rodale at increasing the organic matter of soils using cover crops and manure.

Yet when calculating the ration of emissions to sequestration, it is important to recall that adoption of agroecology would imply breaking the polluting global commodity chains that characterize agrocapitalism, supply lines that contribute to the massive carbon footprint of the industrial food system, which, according to estimates by the nonprofit GRAIN, accounts for between 44 and 57 percent of all greenhouse gas emissions.[110] Agroecology thus promises to significantly diminish greenhouse gas emissions *and simultaneously* revivify soils that are capable of absorbing genuinely significant quantities of the remaining emissions from sources such as industry and transportation. Agroecology could serve as what the ecologist Robert Biel calls a benign form of geoengineering, a low-cost, job-creating, and health-sustaining alternative to unworkable but nonetheless ubiquitous schemes to cope with climate change through for-profit carbon capture and sequestration.[111]

Agrarian social movements like La Via Campesina have embraced agroecology not only because it builds on age-old peasant practices, and not simply because it produces sustainably high yields while conserving the health of soils, crops, fungi, moths, and other species essential to the web of life. In addition, since it does not rely on the costly corporate-owned inputs peddled by agrocapitalist monopolies like Monsanto and Bayer, agroecology maintains the autonomy of agricultural workers from the economically and socially destructive dynamics of agrocapitalism. Agroecology's focus on minimal external inputs, attunement to local environmental conditions, and building biodiversity, among other factors, allows its practitioners to resist capture by key elements of agrocapitalism, including expensive inputs like commodified seeds and chemical pesticides. Simply put, if you do not have to go into debt to pay for seeds and fertilizers, you have a much stronger shot at retaining your autonomy as a producer.

This is the meaning of La Via Campesina's other key term: *food sovereignty*. It alludes to the capacity to evade and overcome agrocapitalism's obliteration of rural society through political mechanisms such as free trade agreements. Agroecology is a method of production that ensures the political self-determination of the peasants, Indigenous people, and landless workers who constitute agrarian social movements such as La Via Campesina. The right to define one's food system is only guaranteed if one controls the means of production. Revolutionary transformation in the agricultural sphere must take place both through cultivation practices and through political organization.

Food sovereignty is not reducible to traditional models of political independence as defined through the nation-state, although it clearly is intended to shore up efforts by Global South nations to fight free trade agreements, damaging claims to intellectual property rights, and other machinations of the neoliberal era. Nor is food sovereignty based on the model of the isolated, domineering individual as theorized in the European liberal philosophical tradition. Instead, the autonomy that undergirds food sovereignty is grounded in the self-generating force or power (*potenza*) that, as theorists of autonomist Marxism like Antonio Negri have argued, characterizes the working class as a whole. In books such as *Marx beyond Marx*, Negri argues that capital does not produce but rather seeks to harness the power of workers. Notwithstanding the apparent might of global elites, that is, *potenza* emerges from human bodies themselves, much like the voice or the capacity to think and communicate.[112] As a result, labor precedes and ultimately eludes capital's efforts to control and discipline its power. In his book *The Progress of This Storm*, Andreas Malm builds on autonomist theory by suggesting that these same arguments may be applied to nature: capital essentially attempts to dominate and exploit the natural world in the same way that it exploits workers. Yet it is never able to completely control the autonomy of either.[113]

The global crisis of soil fertility and the growth of superweeds impervious to herbicides are but two examples of nature's autonomy, unexpected and destructive by-products of agrocapitalism's domineering approach to the natural world.[114] In contrast, agroecological practices advanced in the name of food sovereignty are based on working with rather than attempting to dominate natural processes. The self-organizing characteristics of natural systems, evident in the symbiotic networks linking plants to fungi and bacteria in the soil, therefore have a parallel in the forms of political and social self-organization that sustain food sovereignty. For instance, while agrocapitalism commodifies seeds, small-scale farmers around the world practice seed saving and swapping, thereby establishing networks of mutuality and solidarity grounded in age-old practices linked to the commons: the shared material, intellectual, and emotional resources that capitalism perpetually seeks to enclose.

Food sovereignty is thus predicated on rejecting the forms of global food imperialism that characterized the Green Revolution and other forms of "development," and in their place forging alternative networks grounded in practices of *commoning*, or cooperation. Other related practices of commoning include community-supported small farms; commons regimes in knowledge; building local food sovereignty; and, above all, struggles for land reform. Efforts to take land out of the hands of elites and to redistribute it to the people who actually do the farming were central to revolutionary movements of the twentieth century, figuring prominently in successful revolutions in countries like Mexico (1917), Russia (1917), and China (1947). In the post-1945 period, land reform was a crucial component of decolonization. As we have seen, the Green Revolution was an effort to suppress movements for redistribution and structural change in the countryside, substituting increased production using fossil fuels for practices of intensive cultivation. But recent research has shown that small farms have higher yields than large ones, as well as greater crop diversity and non-crop biodiversity.[115] Small farms constitute the vast majority of the global food system: 84 percent of farms around the world are less than two hectares. Defending these small farms and the people who sustain them is a key component of food sovereignty.

Food sovereignty is what Robert Biel terms an *emergent process*, a series of practices yoking together radical alternatives to dominant agrocapitalist patterns of physical cultivation with radical forms of politics emerging from the autonomous practices of peasants, Indigenous people, landless workers, and small-scale farmers the world over.[116] That is, unlike supposedly new paradigms such as the model of "sustainable intensification" embraced by United Nations organizations like the FAO and the Conference on Trade and Development over the last decade, food sovereignty is based on transforming both the *techniques of production* and the dominant *social conditions of production*.[117] The radical political forms of organization embraced by La Via Campesina and other agrarian social movements would fail without the agroecological practices that empower them, and, conversely, agroecology would quickly be co-opted by agrocapitalism without the radicalism that anchors agroecology in a broader political project to overturn global capitalism.

We cannot consider questions of agroecology and autonomy without highlighting the struggle for female autonomy. It will not be possible to defeat agrocapitalism without empowering women, who are the most marginalized and exploited sector of rural society.[118] If such an assertion seems surprising, it is because women's primary role in agricultural production is systematically rendered invisible. In the United States, the Agricultural Department's Census of Agriculture asserts that only 36 percent of farmers are women, although 56 percent of all farms have at least one female decision-maker.[119] The assumption seems to be that making decisions and engaging in agricultural labor are different things. The FAO publishes similar statistics on women's marginalization in Global South countries, estimating that they constitute just 13 percent of farmers in Brazil, for example.[120] These official statistics imply that rural women's work is a mere aid to men's agricultural labor (just as seems to be the case in the US).[121] Yet as the article "Without Feminism, There Is No Agroecology" argues, on the small-scale family farms that produce 80 percent of the global food supply, women play a key role in food production, from collecting seeds, to preparing land; weeding; food harvesting, processing, and storage; livestock rearing; and fishing.[122] Women in rural areas are also traditionally responsible for household and reproductive labor, meaning that they often spend up to ten hours each day "caring for the nutritional health and well-being of children, families and communities, cleaning and cooking, fetching water, fodder, and fuel."[123] Challenging the systematic invisibilization of women, La Via Campesina argues that women produce between 70 and 80 percent of the food consumed by the poorest families in the world.

Despite their pivotal role in production and reproduction in rural areas around the world, women face a host of social, legal, and cultural constraints that intensify gender discrimination. They have more limited access than men to a wide range of resources, from land, education, and productive and financial resources, to health care and employment opportunities. In addition, women tend to be excluded from decision-making and labor markets. Women's key role in agricultural production receives little to no state support. In addition, they are also subjected to systemic gender discrimination in the domestic sphere, including sexual exploitation, domestic violence, and reduced food intake.[124]

If women's work is largely invisible in agriculture, their traditional and Indigenous knowledge of agriculture—evident, for example, in their practices of seed saving and exchange—is completely disregarded by agrocapitalism. They are among the most vulnerable of groups to the toxic impact of the pesticides and other chemicals routinely employed in agrocapitalism. And they are also among the most affected by the global rash of resource grabbing by investors and private interests. Lastly, women are also the group most frequently subjected to criminalization for their efforts to defend their communities, natural resources, and bodies from extraction and violence.

A feminist agroecology can reverse these forms of systematic gender discrimination, in the process building women's autonomy. A study conducted by La Via Campesina and the National Association of Small Farmers (ANAP) in Cuba, for example, found that conversion from monocultural agriculture to agroecology significantly improved gendered power relations within peasant families.[125] This was possible because all production activities and income generation tend to be concentrated in the hands of men in conventional monocultural systems. Diversifying farms through agroecology, by contrast, introduces other family members to everyday agricultural practices and provides opportunities for income generation for women. By encouraging the use and preservation of local seeds and crop varieties adapted to local climates and traditional knowledge, agroecology helps to empower women, the traditional keepers of seeds and knowledge.

In Rwanda, the Abishyizehamwe women farmers' cooperative set up an early childhood development center, freeing women to engage in agricultural work and community life.[126] The co-op also established community seed banks to store locally adapted indigenous seeds; helped women integrate farm animals into production, providing milk and manure for organic compost; and planted leguminous trees to feed animals and soils. All of this saved women time and helped boost their productivity and economic empowerment.

Supporting women's productive role in rural economies also means eschewing and even smashing agrocapitalism. It means delinking peasant women from dependence on expensive corporate-controlled inputs and neoliberal markets. As the "Without Feminism, There Is No Agroecology" article argues, women "can rely on their own saved seeds, diverse food production and low-input agroecological production methods and, therefore, they can live outside the vicious and unsustainable cycle of loans, expensive inputs, and damaged health from chemical inputs."[127]

Shifting toward agroecology also implies questioning dominant methods for assessing women's labor, which, as we have seen, render women's work invisible. As Vandana Shiva observes, dominant measurements of gross domestic product assume that if you consume what you produce, you don't produce anything—meaning that women's work in the food economy is reduced to zero.[128] Agrarian social movements have not only challenged the conceptual inability of dominant metrics to see women's work, they have also insisted on equality. For example, ANAP in Cuba explicitly states that agroecology must include equal participation for men and women, according to their capacities and conditions, in both work and decision-making on farms. The education program established by the Landless Workers Movement in Brazil, known as Educar, has sought to expose generations of landless settlers and small farmers to principles of agroecology, a program that features equal representation of women and men at all levels of the organization in order to construct what educators call "a more revolutionary feminism."[129] For its part, it has been a constitutive feature of La Via Campesina to insist on gender representation parity at all levels of the organization.

For LVC and allied organizations, the struggle for a new food system involves making all forms of injustice in the existing food system visible, deconstructing those injustices, and in their place building a new solidarity economy where productive and reproductive work is shared equally. This means that the struggle for agroecology and food sovereignty includes the deconstruction of patriarchal family structures, and the reconstruction of social relations based on free and equal relations and shared responsibilities.

## For a Just and Sustainable Global Food System

Agriculture is at a crossroads, faced with two incommensurable futures: a continuation of neoliberal business as usual tied inexorably to greater fragility, crisis, and ecocide, or an embrace of the models of agroecology and food sovereignty for which agrarian social movements have fought for decades. As La Via Campesina puts it in a 2017 report, "The food, climate, environmental, economic and democratic crises that all of humanity is facing show clearly that a transformation of the current agricultural and food model is vital."[130] Agrarian social movements will lead the way as they fight to scale up agroecology.[131] This means prioritizing the resource needs and knowledge of the world's small-scale agroecological farmers. Such support must include termination of public subsidies for industrial agriculture, and a wholesale shift of such support to agroecological practices and research. Transnational agrocapitalist monopolies need to be broken up, a process that must include the smashing of intellectual property rights on seeds.

In place of such increasingly fragile food imperialism, local food systems and territorial markets should be emphasized. And the goals of women's empowerment embodied in the United Nations Convention on the Elimination of All Forms of Discrimination Against Women (CEDAW) must be implemented through a commitment to the role of women as key protagonists in agroecological production and social reproduction. And, lastly, the multiple ecological functions of agriculture—including agroecology's potent potential role in combating climate change—should be emphasized in public policies, research investments, and public finance.

This latter change in values should include the reprioritization of climate adaptation for rural peoples. Measures that climate-focused funding for rural transition should support include the following:

- *Climate subsidies* for rural people who lose their livelihoods and their means of subsistence as a result of climate change;

- *Climate jobs* to provide employment for those who are unable to survive off the land as a result of the climate crisis;

- *Meaningful and appropriate aid* to cope with climate change, including access to appropriate technologies and knowledge of the regenerative agricultural practices that can reverse a century of destructive agrocapitalism; and

- *Genuine land reform* that takes the land out of the hands of wealthy landowners and corporate agriculture and places it in the hands of small farmers, who need good land to survive on.[132]

Who will lead the charge for this transformation? Given the scale of the current crisis, we need states to spark structural transformation in the agricultural sector. But, as we have seen, in recent decades the capitalist state has deepened its alliance with transnational finance capital. Land reform on a national scale has consequently been abandoned, and a global rash of land grabbing by powerful agrocapitalist interests has taken place.[133] LVC concludes that this dynamic has only intensified the need for movements based on environmentalism from below. Creative and disruptive organizing tactics by rural revolutionaries will feature prominently in this fight. As a recent LVC strategy document argues, movements must engage in radical actions that dismantle the status quo in rural areas around the world. Among these, LVC advocates direct actions such as land occupations, marches, protests, and other forms of civil disobedience. In addition, the organization also militates for the transformation of rural production systems along the agroecological lines I have discussed, as well as for the democratization of knowledge and social relations free of classist, racist, and patriarchal logics.[134]

Can grassroots peasant and farmer movements win the struggle for a just and sustainable food system? The odds are daunting, but farmers the world over are mobilizing with great tenacity, certain that what is at stake is not simply their livelihoods and the land they steward but the global food supply. Their movements have proven to be surprisingly powerful. In November 2021, for instance, the massive Indian farmer protest that I described at the outset of this chapter emerged victorious in its struggle against the corporatization of food and agriculture. That month, shortly before the farmer movement was set to commemorate the one-year anniversary of the day when farmers faced water cannons and tear gas as they tried to take their protest to the heart of India's capital city, Prime Minister Narendra Modi announced that he had decided to repeal the three controversial farm laws that had sparked the protest.[135] Farmer organizations and civil society groups around the world watched the Indian farmer protests closely, aware that not just the food security of eight hundred million people in India was at stake, but that defeat of the farmers would constitute a dangerous intensification of agrocapitalism. Solidarity statements from farmer organizations like the Family Farm Action Alliance and the US Food Sovereignty Alliance called for governments to ensure fair prices for farmers, thereby supporting independent family farmers and localized food systems.[136]

The Indian farmer movement and the many groups and organizations that expressed solidarity with the protesters outside Delhi are part of a broader wave of agrarian social movements making demands on capitalist states while simultaneously asserting their autonomy from the state. They share La Via Campesina's vision of a deepening of peasant protagonism and autonomy, grounded in intertwined struggles for agroecology and food sovereignty. This global movement of farmers insists on the integral role of building social solidarity, including with urban social movements. As LVC puts it, "Only a powerful popular movement, that is both rural and urban, can assure that such a process [of agrarian reform] is carried out."[137] Building such solidarities between popular movements in the countryside and in the cities of the Global South is the great task of this century.

Rural dispossession has produced an unprecedented wave of global urbanization over the last half century. As a result, humanity is now a predominantly urban species, living collectively in places marked by extreme economic, social, and environmental inequalities.[138] Indeed, cities—particularly those in the poorer nations—are on the front lines of the climate crisis. But if life in the world's global cities is precarious, urban social movements are also a site of militant collective action of various kinds. From below-the-radar practices of quotidian survival, to organized mutual aid, to mass uprisings, the city today is a locus of what might be termed *urban climate insurgency*. It is to this growing global climate insurgency in the fast-mutating and increasingly imperiled cities of the Global South that the next chapter turns.

# CHAPTER 2

# URBAN CLIMATE INSURGENCY

On October 6, 2019, the Chilean government announced an increase in peak hour fares, making the Chilean transit system the most expensive in Latin America. A few days later, in response to the fare hike, high school students from the National Institute in the capital city of Santiago organized a "mass evasion" of metro stations.[1] One of the few remaining public schools in the country's largely privatized education system, the National Institute had been a center of protests against neoliberal policies in education earlier in the decade. Brutal police repression of this student movement for public education became a regular feature on Chilean television, conjuring traumatic memories of state repression during the dictatorship of General Augusto Pinochet from 1973 to 1990. In the days after they launched the transit protest, high school students were joined by large groups of college students, who gathered outside metro stations and evaded the fares by jumping over or breaking metro turnstiles.

The "mass evasion" protests quickly turned into a confrontation with the Chilean state. By October 17, metro stations were lined with police, transforming fare evasion into a campaign against the repressive arm of the state. Protesters used social media to coordinate the times and locations of the protests. When police responded by closing the gates outside stations, students and supporters tore down the gates. In many spots, the confrontations escalated quickly, with police employing tear gas to disperse crowds, affecting protesters and ordinary commuters alike. By the afternoon of Friday, October 18, the entire metro system of Santiago had been shut down, forcing commuters to squeeze onto crowded buses or spend hours walking home.

Despite the difficulty people experienced traversing Santiago using a partially incapacitated transit system, most people in the city continued to support the mass evasion campaign. According to the on-the-ground commentary of the historian Romina Green Rioja, the fare evasions kicked the lid off boiling popular discontent in Chile, where social movements had for years been demanding free university education, improved working conditions and labor laws, a halt on the rapid rise in the cost of living, and socialization of the private pension system.[2] This discontent was exemplified in the phrase, "This is not about 30 pesos, but about 30 years," which circulated through social media graphics and graffiti tags around Santiago and other Chilean cities. Opposition to the fare hike reflected broader disgust at and rebellion against the growing social inequality and shredding of the social safety net that resulted from decades of austerity policies. As the police attempted to crack down on fare evaders, people joined in solidarity with protesters either by jumping turnstiles or fighting with the police.

On October 18, Chilean president Sebastián Piñera declared a state of emergency that included a mandatory curfew, to be enforced by the military. The prospect of the return of military forces to city streets revived memories of the mass arrests and killings under the Pinochet regime. People began to mobilize in outrage against the reappearance of the military, which broke the "Never Again" promise that had been central to the return of democracy to the country. Protesters gathered first in small groups in iconic parts of Santiago like the Plaza Italia, but soon more broadly across the city and the nation. By October 21, major unions across the country had begun calling for a general strike, and businesses shut down, either in solidarity with the protests or for fear of the spreading social chaos caused by the president's declaration of a "state of war" in the country.

Since they brought to the surface far broader grievances with the state of Chilean society, the uprisings sparked by the metro fare hike proved hard to quell. Although President Piñera quickly scrapped the fare hike once the rebellion became generalized across Chile, protesters refused to return the country to "normal." The fare uprisings sparked a process of reform and transformation in Chile that is still underway. Perhaps the most exemplary part of this process is the still-unresolved effort to replace Chile's neoliberal constitution, a document drafted and passed under the Pinochet dictatorship, with a more democratic one.

It is worth dwelling on why exactly the fare evasion campaign proved so galvanizing. Unlike other forms of neoliberalism such as the privatization of the country's education system, the fare hike ignited anger across a broad swath of Chilean society. It was a direct assault on people's ability to traverse the city. To crimp or foreclose people's ability to move through urban space, whether to go to work, see a doctor, or meet with friends, is to directly attack a vital element of the urban body politic. There can be no right to the city without a corresponding right to urban mobility.

The mass evasion campaign was also implicitly a fight for climate justice. Although the protests were not necessarily articulated in these terms, the right to affordable, convenient public transit is a key element in the struggle for low-carbon urban living. To strike for the right to accessible public transit is thus to militate for the sustainable city. It is a demand central to the contemporary fight for environmentalism from below.

The contradictions that sparked the 2019 uprisings in Chilean cities are typical of conditions affecting growing segments of the global population. Particularly among the peoples of the Global South, the worsening climate emergency is driving the proliferation and increasing political prominence of urban uprisings around the world. Examples of urban uprisings over recent years alone are legion, from protests in Ecuador against the rapid removal of oil subsidies to the detriment of the urban poor, to huge marches from the favelas of São Paulo to the state governor's palace by masses of people demanding economic support, to anti-curfew protests by slum dwellers in Nairobi. The uprisings don't solely involve rural people, like the Indian farmers discussed in the last chapter, bringing their plight to the attention of urban elites through various forms of direct action. Global urban uprisings are also driven by masses of urban dwellers whose life conditions are rendered increasingly precarious by the compound character of urban crises: extreme economic and social inequality, waves of austerity and privatization, draconian state repression, and increasingly threatening climatic crises such as extreme heat and dangerous flooding.[3]

The COVID-19 pandemic exacerbated these conditions, generating a ferocious crisis in which the most basic conditions of social reproduction were stripped from increasing numbers of people around the globe. The pandemic played out as an intensification of what Evan Calder Williams has characterized as a "combined and uneven apocalypse." In contemporary late capitalism, Calder Williams argues, we've witnessed a shift from apocalypse as an instant event occurring on a global scale to one unfolding in an archipelago of zones of infernal breakdown.[4] City dwellers, who by definition have been stripped of the relative autonomy from capitalist markets that characterizes the rural peasantry, have proven particularly vulnerable to this combined and uneven apocalypse, as the thin margins that allow them to purchase food are suddenly torn asunder, and bare survival becomes the horizon of everyday life.

As we have already seen in the chapter on agriculture, hunger tends to radicalize people. When large numbers of people are unable to afford adequate supplies of food, they tend to question the legitimacy of government.[5] In many historical cases, they also rise up. As climate change intensifies food insecurity, it becomes a key (if not the sole) ingredient in a planetary urban powder keg. Given the intensifying prospects for combined and uneven apocalypse on an urban scale, the city is a key site for insurrection—and the climate emergency is a constitutive feature of these insurrections, whether this is overtly declared or not.

The human condition is now a predominantly urban condition.[6] While 55 percent of people currently live in cities, by 2050 fully 68 percent of humanity is projected to be urbanized.[7] Most urban dwellers today live not in wealthy cities with abundant, sleek, environmentally friendly infrastructure—think Copenhagen or Singapore, for example—but in vast, impoverished urban peripheries in various conditions of illegal or irregular residence, sprawling out from the urban centers that benefit from their labor and their poverty. Ninety percent of urbanization during this century is slated to take place in Asia and Africa, with a few key countries like India, China, and Nigeria accounting for up to a third of the global increase in city dwellers.

Equally important, one in eight people currently lives in slums, according to the United Nations Human Settlements Program (UN-Habitat).[8]

Given the rapid growth trajectory of cities in the Global South, two billion people are likely to live in slums by 2030. The pejorative connotations of the term *slum* shed light on the implicit assumptions in most discussions of urbanization, which tend to see the cities of the core capitalist nations as exemplary of an urban condition to which all cities should aspire and, ultimately, develop. Against this Eurocentric tendency, ways of being urban and making new urban futures need to be seen as "diverse and ubiquitous," in the words of South African geographer Jennifer Robinson.[9] This chapter aims to contest the parochial bias of dominant urban theory by recentering discussion of the urban condition in the cities of the Global South. I continue to employ the term *slum*, however, since social movements such as Slum Dwellers International—a federation of urban groups organized along lines very similar to La Via Campesina—use it rather than more bureaucratic descriptions like *informal settlement*.

The UN defines a slum household as a group of individuals living under the same roof who lack one or more of the following: durable housing that protects against extreme climate conditions, easy access to safe and affordable water, access to adequate sanitation, and security of tenure that prevents forced eviction.[10] Slum dwellers also typically live in crowded conditions of more than three people to a room. Such conditions are not exclusive to cities of the Global South. When Hurricane Ida hit New York City in 2021, eleven people living in cramped, windowless basement apartments in the borough of Queens drowned.[11] In the fast-growing cities of the Global South, however, the majority of citizens also lack access to adequate transportation, energy systems, and other infrastructure. Furthermore, they cannot take for granted stable employment or access to healthcare or education.

Living in states whose capacity has been shrunk by decades of austerity policies, most slum dwellers can expect little help from the government.[12] As a result, they must construct and maintain the infrastructures necessary for survival themselves. To live in the city therefore means to create forms of autonomous urbanism. Such autonomous building of lifeworlds has been characterized by commentators like the anthropologist James Holston as a form of *urban insurgency*. Urban insurgency is a set of practices through which people relegated to the peripheries of the world's growing megacities organize themselves, transform the fabric of the city, and assert demands for new forms of citizenship based on access to housing, property, basic infrastructure such as plumbing and electricity, daycare, security, and other guarantees of social reproduction.[13] As Holston argues, "In the process of building and defending their residential spaces," these urban citizens "not only construct a vast new city but, on that basis, also propose a city with a different order of citizenship."[14] Urban insurgency, in other words, calls for a radical revision in our ideas of who builds cities because slum residents are actively constructing the urban fabric rather than waiting passively for state aid.

These forms of autonomous urbanism and insurgency are seldom thought of as a species of environmentalism: as this term developed from the mid-twentieth century onward, it tended to be synonymous with the "natural world" in general, and the countryside in particular. This meant that cities and nature were perceived as geographical and cultural opposites, with cities seen as manufactured social creations, and nature as a space outside of human creation.[15] According to this dominant version of environmentalism, you had to venture out of cities to places like national parks in order to encounter the environment.

Yet cities are obviously not separate from the natural world. Indeed, they are perhaps humanity's most consequential effort to shape nature. Cities are at the core of what radical environmental thinkers call the Capitalocene: the era when the capitalist system's frenetic drive toward incessant growth through the accumulation of capital has dramatically destabilized earth systems.[16] According to the United Nations, cities are responsible for up to 75 percent of contemporary carbon emissions and 60 percent of resource use, with transport and buildings being among the largest contributors to greenhouse gases.[17]

Notwithstanding their massive environmental impact, cities and the urban scale have been remarkably invisible in discussions of the climate emergency. Carbon emissions, for example, are most often tracked through national statistics, or in terms of per capita individual emissions that are themselves tabulated based on a nation-state framework. Yet since most of the world's cities are in low-lying coastal areas, urban populations are at the forefront of the climate emergency as seas rise and anthropogenic environmental disasters strike. How we design cities, and how we conceive of what can be termed the *urban metabolism*, has momentous implications.[18] Will cities function as the equivalent of a gaping maw ingesting resources from around the world and spitting them out as refuse, or as a regenerative formation characterized by cyclical flows? If it is to be the latter, if cities are to be just and sustainable, we need to upend current dominant practices of urban design and development.

Nowhere is the problem of urban nature more evident than in what the urbanist Mike Davis dubs "slum ecologies," the perilous urban environments inhabited by the world's insurgent citizenry.[19] Today's movements of autonomous urbanism must grapple with increasingly extreme forms of climate emergency, from spectacular "natural" disasters like typhoons and hurricanes to more slow-moving forms of climate change–related disturbance like urban heat waves. People who inhabit cities in the world's tropical zones, ground zero for the climate emergency, will be forced to adapt to increasingly blistering temperatures in coming decades. They and their children will confront novel pathogens and pandemics, with immune systems weakened by the environmentally stressed conditions in which urban squatters are forced to settle.[20] Perhaps most importantly, residents of the world's megacities will struggle to feed themselves as the world's agricultural systems are placed under increasing strain by the climate emergency. These challenges, perhaps the most pressing ones collectively faced by humanity today, should be seen as the foundation for an increasingly important global phenomenon: the urban climate insurgency.[21]

Urban climate insurgencies are the responses of global majorities, today's equivalent of the people whom Frantz Fanon called "the wretched of the earth," to the increasingly extreme conditions of the contemporary urban order.[22] It is important to understand that the climate emergency forms a political unconscious that is a constitutive feature of all contemporary public events. Urban climate insurgencies, in other words, don't just include uprisings that explicitly challenge planetary ecocide; they extend all the way to struggles to guarantee social reproduction in the city. The urban climate insurgency should consequently be seen as an inherently intersectional form of mobilization.

This chapter explores the evolving character of urban climate insurgency. In keeping with the notion of a spectrum of intersectional struggles, the chapter explores both social movements engaged in the autonomous construction of urban infrastructures, from housing to transportation and green space, and movements that challenge the state and elites on the conditions of life in contemporary cities. It should not just be rich people, cars, and pavement that have a right to the city. Housing should not just be a human right: the billions of people living in precarious conditions in cities around the world should have a right to a *good* house—a house with excellent insulation, roofs with embedded solar energy arrays, solar water heaters, and rainwater collection systems. To make these homes truly sustainable, they would need to be linked to the rest of the city by high-quality, affordable, low-carbon forms of public transportation. Green homes also need to be braided into verdant public spaces that allow the city to breathe, spaces that give life to a sense of collective belonging and communal recreation. In sum, an ecological reconstruction could regenerate Global South cities, going beyond simply upgrading the conditions in which the global majority dwells to ensure that human habitation in these places is both low-carbon and adapted to the storms to come. In the process, millions of good climate jobs would be generated, helping to lift the world's city dwellers out of the economic precarity that is an endemic feature of today's extreme cities.

Who will build these gleaming green cities? Or, as the great radical urbanist Mike Davis puts it in a seminal essay, the question of who will be the protagonists of low-carbon urbanization in the Global South is perhaps the most momentous question facing the planet.[23] As Davis demonstrates in his often-apocalyptic book *Planet of Slums*, the vast informal settlements that blanket cities in the Global South are, to a significant extent, a result of austerity policies imposed by Global North nations and the powerful international financial institutions, like the World Bank, that they control.[24] These policies have left many nations and metropolitan authorities in the Global South with little capital to devote to slum upgrading. Such policies and the odious debts that underpin them must not only be wiped way: Global North nations that have benefited from centuries of imperial plunder and extraction need to pay reparations for these destructive policies.

Writing on the website of the organization Slum Dwellers International, activists Sheela Patel and Sohanur Rahman argue, "The injustice of climate impacts means the strongest resilience—'survival resilience' built on compound crises—is developed by the world's poorest communities. It is often informal and deeply local. Crucially, it is not fixed or static, due to the unpredictability of climate change impacts."[25] As we will see, the autonomous construction of contemporary cities is a vital component of environmentalism from below, and the best chance for winning the fight against urban injustice and climate chaos.

## Invisible Cities

An urban planner is dispatched to survey a squatter community located on the fringes of a city on a tropical island that once generated immense riches for the French empire. The shantytown is known as Texaco since it is located on polluted land once used by an American oil company. Now that the community has been connected to the rest of the city by a road, it has become attractive real estate for city elites. The city council has sent the planner, who is nicknamed Christ by the people of Texaco, to coordinate expulsions and relocations before the shantytown is demolished. Someone in the community lobs a stone at him, perhaps seeing him—not without reason—as an emissary of an agency bent on destroying poor quarters of the city, "civilizing" their inhabitants by expelling them into bleak housing projects for the poor. Community members take the prostrate young man to the home of Marie-Sophie Laborieux, one of the founders of their shantytown.

Realizing the planner's importance to the community's future, Laborieux sets out to narrate the shantytown's history in an attempt to win his sympathy and thereby save her community. Her story is the center of the Martinican author Patrick Chamoiseau's novel *Texaco*. The narrative that Laborieux tells as she nurses the urban planner back to health depicts the epic battles she and her ancestors have fought to gain a home in the highly cosmopolitan but also extremely polarized social geography of Fort-de-France, the capital of the Caribbean island of Martinique.

Laborieux's efforts to open the eyes of the urban planner to her community's complex history and tenacious fight to survive in the interstices of the city is a powerful evocation of resistance to the forces aligned against slums the world over. Dominant urban theory, together with urban planners such as the unnamed character in *Texaco* who implement such theory, effectively erases these settlements and their denizens. As the geographer Jennifer Robinson puts it, "Theories of modernity have often reserved the experience of dynamism and innovation for a privileged few, and especially for those wealthy cities and citizens who have laid claim to originating modernity. In the process, poorer cities and marginal citizens have been profoundly excluded from the theoretical imaginary of urban modernity."[26] Such cities and their inhabitants are, quite literally, made invisible by urban planning.

Even though the majority of the world's city dwellers now reside in cities in the Global South, these cities tend to be represented as backward and in need of "development." Burdened with a reputation for being outdated, such cities and their occupants are seen in terms of "lack": lack of adequate housing, lack of sanitary infrastructure, and even lack of the cultural forms of "urban modernity" always tacitly associated with Global North cities. As Robinson rightly stresses, dominant urban theory thus reproduces the binary distinctions between the West and the rest that, as scholars such as Edward Said argued, characterized colonial discourse.[27] Such discourses have stark material implications. As *Texaco* demonstrates, once the lives of people like Marie-Sophie Laborieux are made invisible, their place in the city—their homes, their community, and their lives— are all too easily expunged.

In criticizing dominant urban discourse, I certainly do not intend to romanticize subaltern urbanism. There is a long and deeply problematic history of casting slums as sites of plucky bootstrapping, where the poor fend for themselves far better than the state ever could. This approach dovetails all too neatly with neoliberal framings of squatters as entrepreneurial subjects whose shacks constitute a new frontier for surplus accumulation.[28] Contrary to such narratives of self-empowerment, a majority of people living in cities are forced to endure conditions of extreme material and psychic precarity.[29]

Housing is no longer for people: it has become an asset for predatory equity funds that rove the globe in search of "undervalued" investment opportunities. During the neoliberal era, housing has been turned into one of the world's prime commodities. As geographer Samuel Stein documents in *Capital City*, global real estate is now worth more than $219 trillion.[30] Over 60 percent of the world's financial assets are in real estate, with the majority of that figure in housing. In a wealthy country like the United States, over a third of home sales were to absentee investors in 2016. But the problem is a global one: as a result of the financialization of housing, record numbers of people worldwide have been rendered homeless in what sociologists like Saskia Sassen and Matthew Desmond describe as a process of brutal expulsion.[31]

The critical literature focused on evictions taking place in the US and other wealthy nations roots displacement in market dynamics, which tends to individualize the problem, as if it were simply the improvident behavior of specific people that is to blame for the housing crisis. But forced removals of informal, autonomously constructed slum settlements in the Global South show that evictions are political events that affect entire communities rather than isolated individuals or families.[32] The strong arm of the state is almost always in the mix, even when it more or less openly serves the interests of private developers. Forced removals, or what anthropologist Arjun Appadurai calls "urban cleansing," reanimate logics of exclusion left over from the colonial construction of many cities.[33] Residents respond collectively to the threat of forced removal using a wide repertoire of resistance strategies that also have roots in anti-colonial resistance.

In the teeth of these hostile conditions, squatters the world over often find ways to assert themselves, to establish makeshift forms of community, and to remake the city through forms of autonomous political organizing that are often radical and militant.[34] Rather than seeing Global South cities as simply *lacking* certain infrastructures, we must instead see how cities are created and sustained through acts of urban bricolage—a contingent and "below the radar" art of social assembly that can lead to bottom-up material transformation, in contrast to the top-down, perfectly executed master schemes of architecture and urban planning. Instead of an absence of infrastructure, Global South cities must instead be seen as characterized by an abundance of "people as infrastructure," a phrasing coined by the urbanist AbdouMaliq Simone.[35] The acts of bricolage through which improvised urban lives are assembled can often suggest a political openness, a willingness to consider and fight for alternative urban futures.

The right to inhabit a city is extremely tenuous for vast numbers of people. Globally, the gap between the amount of affordable housing and the number of people who need it is increasing rapidly. By 2025, this shortfall is expected to affect over a third of the world's urban population, meaning that at least 1.6 billion people will struggle to survive in unsafe and overcrowded high rises or informal slums, or with no lodging at all, sleeping rough on pavements and under highway overpasses.[36] Processes of class-based dispossession play out in cities across the Global South through what the geographer Tom Gillespie has termed "accumulation by urban dispossession."[37] In his account of evictions in Accra, Ghana, Gillespie argues that members of the city's large informal proletariat, people excluded from formal wage labor and housing markets, must create forms of urban commons in order to survive in the city. But since these commons place limits on capitalism's capacity to generate profits through the development of urban real estate, state-led campaigns of urban dispossession target the informal poor for expulsion.

Alongside brute force and legal chicanery, elites use revanchist discourses to legitimate expulsions. As Gillespie argues, they frame squatters as "a source of dirt and disorder whose presence is undermining efforts to transform" the city into a place friendly to local and international capital.[38] The continuity between colonial-era sanitary discourses and *cordon sanitaire* strategies and the discourses that legitimate contemporary expulsions is striking. Often, these ideological legitimations for expulsions describe mass removals as an environmental improvement strategy—despite the relatively low environmental impact of slum dwellers.[39] The link to forms of "green gentrification" in Global North cities is notable.

It is relatively easy to legitimate mass expulsions and slum clearances in this way because the people living in these places do in fact face immense environmental challenges. As Davis puts it in *Planet of Slums*:

> The cities of the future, rather than being made out of glass as envisioned by earlier generations of urbanists, are instead largely constructed out of crude brick, straw, recycled plastic, cement blocks, and scrap wood. Instead of cities of light, soaring toward heaven, much of the twenty-first century urban world squats in squalor, surrounded by pollution, excrement, and decay.[40]

Urban life for the global majority is by definition hazardous and health threatening.[41] Davis rightly argues that cities in the abstract are a potential solution to the global environmental crisis: urban density can promote the construction of more efficient infrastructure, including social housing and mass transit, which results in low per capita carbon emissions. In addition, cities can create the kinds of public spaces and forms of democratic governance that are essential to planning for the common good. But such urban environmental efficiency and public affluence require preservation of what Davis calls a "green matrix" of intact ecosystems, open spaces, and natural services.[42] Tragically, elites in many cities of the Global South are razing and polluting crucial environmental support systems such as wetlands and forests.

If cities are developing in environmentally dysfunctional ways, urban squatter communities often bear the brunt of such maldevelopment, since squatters generally seize derelict land in the most environmentally fragile zones of the city. From the barrio built on stilts above the excrement-clogged Pasig River in Manila, to Mumbai's Dharavi slum, which sits on a mangrove swamp in the heart of the city, slum dwellers are threatened both by the state because of their "illegal" settlement on disused public land, and by the environmental vulnerabilities that result from settling in such places. As the planet heats up, it is slum dwellers who will be the most exposed to deadly heat increases. In 2017, researchers in Hawaii projected that if greenhouse gas emissions continue to grow, the share of the world's population exposed to extreme heat for at least twenty days a year will increase from 30 percent now to 74 percent by 2100.[43] If there is a front line in the climate emergency, the denizens of what urban researcher Robert Neuwirth calls "shadow cities" are living on it.[44]

But the gamut of environmental hazards to which slum dwellers are exposed is not simply a result of their occupation of vulnerable urban zones. Most often, it is a result of social engineering and, more specifically, of histories of racism and ongoing forms of inequality. In addition to coping with flooding and extreme heat, slum dwellers are faced with myriad forms of life-threatening pollution resulting from the toxic legacies of colonial and postcolonial planning. In constructing cities across Africa, Asia, and the Americas, European imperial powers refused to build modern sanitation and water infrastructures in neighborhoods outside the pale of European urban settlement. Colonial powers chose instead to use racial zoning to segregate white quarters of the city from the epidemic diseases produced in "native quarters" of the city by the lack of sanitary infrastructures.[45] As Frantz Fanon documented in scathing detail in *The Wretched of the Earth*, colonial cities were spaces in which inequality and injustice were manifest in the most palpable and enraging of ways:

> The settlers' town is a strongly built town, all made of stone and steel. . . . The settler's town is a well-fed town, an easygoing town; its belly is always full of good things. The settlers' town is a town of white people, of foreigners.
>
> The town belonging to the colonized people, or at least the native town, the Negro village, the medina, the reservation, is a place of ill fame. . . . It is a world without spaciousness; men live there on top of each other, and their huts are built one on top of the other. The native town is a hungry town, starved of bread, of meat, of shoes, of coal, of light. The native town is a crouching village, a town on its knees, a town wallowing in the mire.[46]

The colonial city was a machine designed to ostracize the colonized. Fanon notes that this rejection comes not simply through the material impoverishment endured by denizens of the colonial city, the overcrowding and unsanitary conditions they must survive. The city also heaps humiliation on the colonized, making them feel completely abject, and thereby producing a town that wallows in the mire both in real material terms and in emotional terms. The colonial city was a city of absence, a place consciously deprived of electricity, food, and clean water, a site of purposeful negation.

Since this lack of infrastructure was not systematically addressed in most postcolonial cities, where governing elites often simply moved into the former European colonial districts of the city, illnesses related to the lack of sanitary infrastructure kill shocking numbers of people in the cities of the Global South. According to the World Health Organization, at least two billion people around the world are forced to use water sources contaminated with feces; such contaminated drinking water is estimated to kill over 1,300 people every day.[47] The majority of the people who die are children.

Neoliberal policies of austerity imposed by international agencies during the last four decades have intensified the toxic legacy of colonial urban inequality. For decades the IMF and World Bank have advocated regressive user fees and charges for public services, while never proposing taxes on the wealthy, conspicuous consumption, or real estate barons.[48] But urban elites and the state in postcolonial nations are also responsible for the appalling conditions in Global South cities, including the lack of housing and essential infrastructure. In the Mumbai region, for example, urban development authorities have effectively confiscated local power, ensuring that modern infrastructure is built only in wealthier neighborhoods.[49] The betrayal of promises to the poor and the gutting of urban democracy by local and global elites has generated a situation in which, as Davis puts it, "excremental surplus is the primordial urban condition."[50]

Urban dwellers are also menaced by airborne toxicity. The combination of sprawling urban growth with anemic investment in public transit has turned traffic into a major public health crisis. Globally, air pollution accounts for about seven million premature deaths a year, according to the World Health Organization—although recent research puts pollution's toll far higher than that estimated by the WHO.[51] The majority of those deaths are the product of outdoor air pollution, although smoke produced by indoor cookstoves in cities where people lack electricity is also an important factor. Most of these deaths occur in developing countries, with China and India accounting for half of the total, although air pollution also remains a significant killer in cities in the Global North as well. India is home to nine of the ten most polluted cities on the planet, where city dwellers find themselves cloaked in choking clouds of smog. There are quite a few overlapping factors to blame for this death-dealing air pollution, including trash fires used to burn uncollected garbage, the use of dirty diesel generators during frequent power outages, and the clouds of smoke that drift over cities when farmers set fire to their fields to clear them after harvest. But growing pollution produced by transportation is an undeniably prominent contributor to bad air.

The pollution problem is getting much worse. While some cities in the Global North may appear to be going greener through innovations like bike lanes and new public transit systems, such gains in sustainable transit systems are more than offset by the rise of fossil fuel–based transport and emissions on a global scale. In its *Global Mobility Report 2017*, the World Bank warned that passenger traffic is set to increase 50 percent by 2030, with the number of cars on the road doubling to nearly 2.5 billion, while global freight volumes will increase by 70 percent.[52] Increasing automobile traffic is also leading to growing carbon emissions. Transport is already the leading source of greenhouse gas emissions in developed nations like the US and UK, but global land transport emissions could grow to 13 gigatons per year by 2050 as a result of a near-tripling of transport emissions in Global South nations.[53] Although transport currently accounts for only 14 percent of total global emissions from human sources, it is particularly heavily dependent on fossil fuels—constituting, for example, two-thirds of oil demand, according to the International Energy Agency.[54] Growth of fossil-based transit in the Global South will consequently significantly increase carbon emissions. In addition to killing people and the planet slowly, cars also kill people swiftly, with traffic accidents taking the lives of more than one million people each year in the Global South.

As is true for waterborne mortality, all of this mass death—which might best be termed *environmental necropolitics*—is no freak of nature.[55] It is a product of carefully constructed policies. For instance, austerity policies adopted by international agencies like the International Monetary Fund have played a key role in spreading the modes of transit that generate deadly air pollution. Decades of austerity policies emanating from these international institutions have starved public transit of funds, with resulting problems being used to legitimate further cuts. Direct funding has also been provided to the private sector by agencies like the IMF, since this can be kept off a government's account books while funding for public transit must be put on the public ledger. The result has been systematic support for climate-killing forms of transit like highways. As labor activists Sean Sweeney and John Treat point out in their report on climate-ready mobility, global levels of investment in public transport have shrunk, while the money directed to transport infrastructure has gone mostly to accommodating more vehicle travel through road building, to facilitating international trade, and to publicity-grabbing megaprojects that serve the business class, such as the construction of Cambodia's $1 billion New Siem Reap-Angkor Airport.[56] As a result, aviation emissions are set to grow 300–700 percent by 2050, according to the International Civil Aviation Organization.[57] So preference goes to building roads, ports, and airports rather than public buses and rail transport in most developing countries, even though the latter are far more efficient, both in terms of their low-carbon footprint and in terms of moving masses of people around. Policies of urban maldevelopment marginalize urban majorities in the Global South because these communities seldom can afford the forms of private transit supported by international agencies.

These policies are part of the longer history of rendering the urban subaltern invisible, feeding into contemporary accumulation by urban dispossession.[58] Yet while contemporary theories of subaltern urbanism have often championed popular agency, they seldom conceive of these acts of resistance in environmental terms. But urban squatters clearly transform their environments, often in the face of some of the world's biggest contemporary environmental challenges. The next section elaborates on some of the grassroots environmental solutions developed by urban climate insurgencies in various cities around the world. Contrary to tendencies in the Global North to see environmental struggles unfolding on the terrain of some non-urban, non-anthropogenic thing called "nature," urban environmentalism from below transforms the built world and infrastructural systems, thereby intertwining struggles for social and climate justice.

## Urban Environmentalism from Below

In Chris Abani's novel *Graceland*, when the Nigerian government sends troops to clear residents out of the swamp city of Maroko through what it calls "Operation Clean the Nation," slum dwellers do not talk. They organize. They fight. And some of them die rather than leave the homes they have built. The squatter settlement of Maroko, which was a real community in the early 1990s, was located near the heart of downtown Lagos, just opposite the chic neighborhoods of Victoria Island and Ikoyi.[59] Prime real estate, it came to be coveted by the Nigerian elite and foreign investors, and so the squatters who had built precarious shacks on stilts above the swampy land had to go.[60] Their eviction forms the culmination of *Graceland*, which memorializes the demolition of Maroko in 1990 by the military junta of Lagos governor Colonel Raji Rasaki as part of the government's "War Against Filth."[61] Three hundred thousand residents were ejected from Maroko after colonial-era legal statutes marked as illegal the informal dwellings in which the majority of the city's denizens lived. This striking continuity in strategy and rhetoric across the colonial and postcolonial eras painted poor citizens as unsanitary elements in the body politic, pollutants that in the name of national hygiene had to be evacuated using martial law and militarized violence.[62] Abani's novel focuses on the Maroko residents' resistance to this eviction. Through the medium of fiction, Abani alerts readers to the fear of displacement, the boiling rancor, and moral outrage that animates resistance to eviction the world over.

An article in a Nigerian publication revisiting the demolition of Maroko on the twenty-fifth anniversary notes the violence with which residents were expelled from the community by the police and military, who stole and looted belongings, killed residents who got in their way, and also allegedly raped women and children.[63] Evictees from Maroko were transferred to two abandoned government estates many kilometers away from their demolished homes. These sites had no electricity, roads, or running water. According to Maroko residents who were moved to these distant sites, children in the community suffered separation from family members, leading to feelings of "desertion, loneliness, hopelessness" as well as hunger and illness.[64] These specific experiences illuminate broader global dynamics where the clearance of coveted land in strategic parts of cities leads to dramatic deteriorations of displaced people's economic prospects, health, and psychological welfare.

The injustice of slum clearance for speculative development is stark: real estate bubbles ensure that significant numbers of houses lie vacant in cities—and not just rich ones—around the world. Meanwhile, although the proportion of people living in slums in the Global South decreased in the period from 1990 to 2014, absolute numbers of people living in such conditions rose by 28 percent.[65] Although exact numbers are hard to pin down, by any measure the housing crisis is acute, with an estimated 1.6 billion people living in informal settlements. The number is set to swell to

2 billion by 2030. In some cities, proportions of people living in slums may rise to as high as 90 percent.

The resistance campaign that developed against the impending demolition of Maroko challenges the still-widespread belief that slum dwellers are incapable of organizing themselves and asserting their collective agency.[66] Abani's novel lays out some of the strategies that residents of slums deploy to resist displacement, including collective planning, outreach efforts to the media, and, ultimately, physical resistance to the police demolition operation. This fictional account of the resistance in Maroko aligns with the strategies documented by scholars and activists in cities around the world.[67] Key strategies identified by a global network of grassroots organizations fighting eviction include negotiation with public authorities, public campaigning for recognition of rights, legal challenges to evictions, and, finally, open struggle against state authorities engaged in violent evictions.[68] Resistance tactics differ considerably in different urban and national contexts, yet in each place squatters display a remarkably tenacious and creative refusal to go along with state-sanctioned displacement policies.

In the South African city of Durban, for example, the organization Abahlali baseMjondolo (which means "shack dwellers" in isiZulu) formed in 2005 in response to broken promises by local politicians to improve the conditions in the city's informal settlements. The group blocked a major road for hours to draw attention to their situation, and then organized a huge protest march after police cleared the demonstration by force. The group fought against arrests in the country's courts while also simultaneously mobilizing in the streets to demonstrate people's power. Abahlali baseMjondolo's key demand for an end to forced removals is part of a broader series of struggles for access to education and the provision of water, electricity, health care, and sanitation that together amount to a form of bottom-up democracy.[69]

Slum dwellers around the world fight eviction because they understand all too well that policies of forced relocation to the urban periphery cut them off from access to the social networks and jobs on which they depend. People don't want such houses, even when they're on offer for free. For instance, an assessment of Mexico's Iniciamos Tu Casa (We Start Your House) program, which provided poor residents with housing outside the city, found that many participants abandoned their houses within a year after the program started.[70] Although these home-building and relocation policies have proven to be a failure, governments continue to pursue them, using names that clearly advertise the ideological linkage between home ownership and social pacification, from Angola's My Dream, My Home program to Brazil's Minha Casa, Minha Vida (My Home, My Life).[71] But social connection is worth more than simply having a government-constructed roof over your head. Despite these home-building policies, communities prefer access to services, jobs, and the less tangible social networks that the urbanist AbdouMaliq Simone calls "people as infrastructure."[72]

Poor people's resistance to displacement is also an unacknowledged form of grassroots environmentalism.[73] By taking a defiant stance to remain in their homes, slum dwellers mitigate against urban sprawl and defend relatively low-carbon forms of habitation. Eviction to the distant urban periphery cuts communities off from existing transportation networks. The commutes from distant settlements to urban centers where jobs can be found are not just gruelingly long and costly; they are also energy intensive. The policies that displace people toward the urban periphery intensify ecologically unsustainable patterns of sprawl, which restrict future efforts to build urban resiliency and lock in pathways of high-energy use and carbon emissions.[74]

Settler-colonial nations like the United States, Australia, and South Africa may have established this paradigm of environmentally dysfunctional suburban sprawl (often for blatantly racist reasons), but much of the rest of the world has now begun to emulate this model, building cities characterized by ever more disconnected peri-urban growth.[75] As a result, cities are growing increasingly congested and polluted. In addition, as we have seen, transportation is becoming a more significant contributor to greenhouse gas emissions. Low-carbon cities are compact cities, and poor people are adept at building precisely such urban forms.[76] Squatters, in sum, are urban climate insurgents.

There is a lengthy tradition of celebrating the autonomous community-building activity of slum dwellers. In 1968, the British architect John Turner set out to challenge negative depictions of squatters in an article describing what he called "the architecture of democracy" in Lima, Peru.[77]

Despite the negative associations enduringly attached to slums, according to Turner they exemplify the maxim that "the man who would be free must build his own life." In most instances, he insisted, squatters were building their own homes as part of an escape from paying rent to slum-lords, and were thus "consolidating an improved status."[78] Turner argued that a squatter-built house was a step up from renting a clandestine shack from someone else, and that such autonomously constructed housing was the product of three freedoms: the freedom of community self-selection, the freedom to budget one's own resources, and the freedom to shape one's own environment.[79]

The first of these freedoms is fairly self-explanatory: slum dwellers decide where they will live, albeit not under conditions of their own choosing. This tends to generate a sense of agency and collective unity. Because slum associations, groups through which slum dwellers make collective decisions about issues such as the construction of infrastructure, generally offer inexpensive membership dues, the bar to acquiring a plot tends to be quite low. In contrast, housing provided by the government or private builders typically requires a large initial outlay of capital. But the slum dweller can take up residence for a negligible sum, and then gradually improve their habitation as circumstances allow. This is why Turner argued that squatters have freedom to budget their own resources, and consequently to embody a kind of incremental urbanism.

Lastly, Turner pointed out that housing built by the government or private interests imposes on the poor an authoritarian structuring of space and personal relationships. Turner argued that squatters who build their own homes, on the other hand, are able to adapt their space based on the changing needs and behavior patterns of the individual or family. They therefore have freedom to shape their own environments. Although Turner did not discuss squatters' behavior in these terms, the last freedom he proposed was also remarkably prefigurative. This freedom, understood today, should also be seen as encompassing the ability to shape the built environment in response to local manifestations of the climate crisis. After all, who knows more about the specific challenges and necessary solutions to environmental breakdown than those who live on the front lines of the climate crisis?

Turner was writing in the context of an anti-authoritarian moment in urbanism and in intellectual culture more broadly. This was the era, as architectural critic Justin McGuirk notes, when Jane Jacobs defended the boisterous streets of Greenwich Village against Robert Moses's community-crushing highway construction, and when the radical Situationists in France pilloried master architect Le Corbusier and his tower-bloc "morgues."[80] All too quickly, however, Turner's celebration of the existential empowerment of slum dwellers was appropriated by powerful global financial institutions, whose neoliberal policy prescriptions were to have a profound impact on cities around the world for the last quarter of the twentieth century. Celebration of squatters' self-built housing has come to seem complicit with this legacy of state abandonment. In scathing terms, for example, Mike Davis describes the "intellectual marriage" between Turner's paean to spontaneous urbanization and World Bank president Robert McNamara's austerity agenda. As Davis puts it, "Amidst the ballyhoo about 'helping the poor help themselves,' little notice was taken publicly of the momentous downsizing of entitlement implicit in the World Bank's canonization of slum housing. Praising the praxis of the poor became a smokescreen for reneging on historic state commitments to relieve poverty and homelessness."[81]

Using the intellectual ammunition provided by Turner and other advocates of self-help in the cities of the Global South, the World Bank's Urban Development Department funded the implementation of a new set of policies that rejected state construction of public housing in favor of what it called "sites and services" projects. In such schemes, funding was provided solely for a plot of land and basic infrastructure such as water and electricity; poor people in cities from Manila to Mumbai were then expected to build their houses themselves. This obviously saved governments a huge amount of money. It is hardly a coincidence that international financial institutions were at this time also pressuring nations of the Global South to pay back the external debts they had accrued when they were encouraged in the 1970s and 1980s to borrow capital for development projects from banks in wealthy countries. Site-and-services projects can thus be seen as one set of programs within a much broader realignment of the state in postcolonial nations through austerity measures that came to be known as "structural adjustment programs." This was imperialism working its way through the urban fabric.

Yet by the time Davis published his stinging critique of neoliberal urban programs in *Planet of Slums*, the World Bank—reacting to similar earlier criticisms—had already wound down its site-and-services program. What was the alternative? As Justin McGuirk documents, if Latin America was the first region of the Global South to experience mass urbanization, it was also the place where state-provided mass housing programs failed. Even under the region's military dictatorships of the 1960s and 1970s, with their corrupt ties to construction firms, the state simply couldn't keep up with popular demand for housing. In addition, the assumptions underpinning the state-led mass housing programs of the post-1945 era were fatally flawed. As McGuirk puts it, "Paternalistic attempts to impose order on the city in the mid-20th century failed because the slums are too dynamic a force in the city to be subsumed and were consistently underestimated by both planners and architects."[82]

Mass housing provided by the private sector, which neoliberal ideologues subsequently argued was going to solve the housing crisis, also proved to be a failure.[83] Not only have insufficient units been built by private developers in cities around the world, but, like government-built units, they tend to be located in the distant outskirts of the city. In addition, bureaucratic hurdles mean that the poorest citizens, and women in particular, often don't qualify for such housing. And as John Turner argued, mass housing seldom fits the specific needs of particular families, and leaves little room for families to adapt space to their needs. So Davis's question regarding who will build the homes and communities that will house the world's urban denizens—and in the process help them adapt to climate change—remains open.[84]

At present the vast majority of housing construction and defense in the cities of the Global South is carried out autonomously by communities, families, and individuals. But what might it look like if urban social movements could claw back some of the power lost during decades of austerity, thereby forcing the state to collaborate in urban environmental reconstruction? The Empower Shack project in a township on the outskirts of the South African city of Cape Town offers a provisional answer to this, one of the greatest questions of this century. Cape Town is one of the most economically dynamic cities in the region, yet approximately 20 percent of the city's households live in informal dwellings—7 percent in informal backyard structures and 13 percent in informal settlements.[85] The post-apartheid city government aims to provide free basic services (water, sanitation, and electricity) to existing informal settlements, but there is still a massive backlog in the provision of government-built affordable housing to the city's poorest residents. Of the estimated more than 2.5 million homes that need to be built across the country, half a million are within the Western Cape Province and three hundred thousand in the city of Cape Town alone. The Empower Shack is an interdisciplinary housing development designed to work as a prototype solution to this housing crisis.

The project unfolded in the mid-2010s through a fourfold collaboration that brought together the architectural group Urban-Think Tank (U-TT), the South African nonprofit organization Ikhayalami, a community in the township of Khayelitsha, and the Cape Town city government.[86] Absolutely key to the success of the Empower Shack project, given the history of forced removals carried out under apartheid, was the fact that it rehoused township residents by engaging them in the rebuilding of their existing housing rather than displacing them. As U-TT puts it, "Using the existing footprint of the shack and working closely with residents and city planners, structures are built creating homes, not just houses."[87]

Four key components help make the Empower Shack project an exemplary model for transforming the cities of the Global South: a two-story housing prototype, a participatory spatial planning process, integrated urban systems, and economic solutions.[88] These elements were realized through generative collaboration between various organizations: Municipal authorities provided basic infrastructure, including water, sanitation, and road infrastructure for a block of fifty or so homes, while the nonprofit Ikhayalami was responsible for organizing and adequately involving all community members in the project. U-TT, for its part, developed "an open-source housing prototype and urban plan" that it envisaged as a "model for informal settlement upgrading across South Africa."[89] The project transformed the existing one-story, corrugated iron shacks that are common throughout South African townships into two-story concrete dwellings that offered residents far more space, on the same or even smaller footprint. These houses are designed to be built in just one day, their facades painted in distinctive colors intended to help foster a sense of pride and community engagement.

The new dwellings were also designed to be flexible and economically inclusive. For example, Empower Shacks are clustered around interior courtyards filled with trees and playgrounds for children, thereby creating precious public space and the social connections that it can help foster. U-TT's design allowed residents to add balconies that looked down on these public spaces, as well as covered, semi-private spaces where residents could plant small urban farms. Included in the project were livelihood programs such as construction apprenticeships, as well as community urban management programs. These public components were designed to ensure that community members were not consulted in a superficial manner about the project, but could actually build the houses themselves, deriving skills and income in the process, as well as an all-important if relatively intangible sense of collective ownership and civic pride.

The Empower Shack project is also strikingly green. The scheme incorporates many elements designed to help promote community adaptation to the climate crisis, as well as design features that simultaneously promote community autonomy and low-carbon living. Among the features that are explicitly designed to lighten the environmental footprint of the homes are roofs clad with solar arrays so that the new homes generate their own electricity. In addition, the complex uses a greywater recycling system, and also features tree-lined streets designed to cool the neighborhood. The solar panels and water recycling components are the visibly green features, but the project incorporates many less obvious but equally important green elements. For example, corrugated iron structures that exposed residents to the Cape's searing hot summers and chilly, rain-filled winters were replaced with structurally sound and relatively insulated permanent dwellings—a key element of the project's environmental upgrading. In addition, rather than moving residents to ever more distant urban peripheries, the project permits communities to relocate in place, thereby helping to ensure a denser and less resource-intensive urban fabric.

Lastly, by engaging the municipality in a collaborative project with community members and a grassroots NGO, the Empower Shack project helps provide residents with security of land tenure. As the urbanists José Rafael Núñez Collado and Han-Hsiang Wang point out, security of land tenure is a key but rarely acknowledged environmental element in slum-upgrading efforts: "When climate events warnings are given, many low-income households will not move or seek shelter elsewhere, despite the risks, thinking that their households are unprotected and worrying that they will not be allowed back. By having legal or de facto ownership of their properties, slum dwellers are more willing to evacuate and to invest in strengthening their structures."[90] Security of land tenure does not necessarily mean land titling, a controversial process that—absent other reforms—can quickly lead to dispossession of newly minted owners who are forced to sell their homes to pay for other necessities. Secure tenure can simply involve the legitimacy that slum-upgrading projects confer on informal settlements in the eyes of state authorities. As Collado and Wang emphasize, such security helps build the trust needed for participation in everything from adaptation efforts to disaster preparation.

Empower Shack builds on the best practices of contemporary slum-upgrading efforts. For example, in its theory of change manifesto, Slum Dwellers International articulates a vision grounded in community-driven slum upgrading: "We see slum settlements that are recognized by the city, have secure tenure, and universal access to basic services. We see safe and healthy communities where social cohesion is evident and even those with low incomes have access to economic livelihood opportunities and pro-poor credit."[91] If site-and-services schemes were predicated on a romanticized notion of the autonomous activities of slum dwellers that justified a lack of state provision, the best contemporary slum-upgrading efforts aim to provide not simply services, infrastructure, and even housing but also seek to integrate informal communities into the life of the city. This requires a shift in perception, eschewing the long-dominant approach of seeing slums as a problem to be ignored or demolished and instead seeing them as vital parts of the city, places whose denizens need to be engaged in the co-creation of solutions to the most pressing challenges as well as longer-term ones like climate change.[92]

As McGuirk documents in his book *Radical Cities*, the most influential examples of slum upgrading have emerged over the last few decades in Latin American countries, places where social democratic governments won power with the support of strong popular social movements, creating generative (if at times tense) exchanges between radicals in the halls of power and movements out in the city streets.[93] In some instances, popular mobilization around housing seems to be surpassing Turner's wildest hopes. In Uruguay, for example, over twenty-five thousand families have organized 560 mutual aid–based housing cooperatives across the country.[94] Organized as the Uruguayan Federation of Mutual-Aid Housing Cooperatives (FUCVAM), this movement promotes what is known as worker *autogestión* (self-management), participatory democracy, and *ayuda mutua*. The latter term is particularly significant, translating not just as "mutual" aid but as the "sweat equity" generated when cooperative members contribute directly by working on building sites for the homes they and other cooperative members will eventually inhabit.

As cooperative members explain in *Sweat Equity*, a film by activist Daniel Chavez about FUCVAM, the movement sees housing as an urban commons. This means that houses built by the cooperative are not privately owned. Cooperative members search for a suitable plot of land for their future homes, and then take out a collective loan from the state to fund construction. As a collective, they assume control of the whole building process and the management of the urban space once construction is completed, a collaborative process that fosters internal solidarity, social empowerment, and democratic innovations along the way. Treating the resulting homes as a commons rather than as privately owned properties also permits a remarkable degree of flexibility over use of space: co-op members initially get access to a two-bedroom house, but they can apply to trade up to a larger home if their needs change over the years.

In existence for four decades, FUCVAM has become an important political force in Uruguay. During the twelve-year period of dictatorship from 1973 to 1985, FUCVAM played a key role in resistance to the regime, organizing massive protest marches and boycotts of key state institutions like the Banco Hipotecario del Uruguay.[95] Although many of the movement's leaders were imprisoned, its wide base of support ensured that it continued to play an important social role in the country. Subsequently, FUCVAM helped stave off neoliberal attempts to privatize public infrastructure such as Uruguay's telecom system.[96] As one of the world's most ambitious and politically radical efforts to solve the housing crisis by mobilizing popular power, the model established by FUCVAM has become an inspiration to movements in other countries in Latin America and beyond.

FUCVAM is one of the most striking exemplars of the broader resurgence of popular movements that characterized the Pink Tide in Latin America.[97] This is the fertile milieu out of which the Urban-Think Tank, the design firm collaborating on the Empower Shack project, emerged: U-TT was co-founded by Venezuela-based architect Alfredo Brillembourg, whose career has been built around improving conditions of informal settlements, such as those where the majority of people in his home city of Caracas live. U-TT's experiments in slum upgrading build on a long history of efforts to tap into the autonomous power of slum dwellers.

For instance, inspired by Turner's ideas about incremental architecture, Peruvian president Fernando Belaúnde in 1968 established the PREVI (Proyecto Experimental de Vivienda) scheme, in which an impressive assortment of renowned architects designed individual houses that families could expand on as they saw fit. Tragically, PREVI's strategy of treating architecture as part of a broader process of social transformation was not replicated for many decades: in the very year that it was established, Belaúnde was overthrown in a military coup. It was not until the return of radical politics to Latin America with the arrival of Pink Tide governments in the late 1990s that conditions would again be favorable for the kind of collaboration between social movements, leftist parties with institutional and economic power, and architects and urban designers that characterizes the best instances of slum upgrading.

One of the first such collaborations was the project of Chilean architect Alejandro Aravena and his firm Elemental. After an earthquake demolished many of the homes in the informal settlement of Quinta Monroy in Iquique on the northern coast of Chile, Aravena and Elemental built an estate of ninety-three houses in the period 2001–4. The homes were constructed at extremely low cost because the firm left them half-completed, allowing residents to move back in and personalize the dwellings by adding additional rooms, verandas, and other elements. Like the abandoned PREVI experiment before it, Elemental's project embodied the community-empowering idea of incremental architecture, but updated it with significant infrastructural and social support from the state. And similarly to U-TT's subsequent Empower Shacks, Elemental's work also incorporated important climate adaptation and mitigation elements. For instance, materials dismantled from the clearance of the original slum in Quinta Monroy were recycled and reused for the expansion of the new houses.[98] As in Cape Town, the new permanent houses were far more structurally sound, and consequently were able to protect residents from the area's temperature extremes, which climate change will certainly intensify. Lastly, the security of tenure that the project conferred on the residents allowed them to invest further in their homes, upgrading them with support from the Elemental team in a manner that ensured efficient use of resources, structure, and materials.

Yet, as important as the Elemental project was as an ecologically sound model of slum upgrading, transforming individual buildings or clusters of buildings will never be adequate to cope with the climate crisis. Designers need to work with social movements in slums to generate networks of vibrant and verdant spaces, and in so doing to transform the entire urban fabric. This more holistic approach to city planning and activism has come to be known as *social urbanism*. Medellín, Colombia, offers one such example. Long synonymous with the hyper-violence unleashed by drug cartels, Medellín underwent a remarkable transformation in the early 2000s based on a series of strategic plans for integrating the informal settlements on the urban periphery into the city's physical and social fabric. Instead of focusing attention and funding on isolated buildings, social urbanism in Medellín was based on the development of what were known as Integrated Urban Projects (Proyectos Urbanos Integrales, or PUIs) that aimed to create what Collado and Wang call "networks of multiple interconnected spaces across a territory."[99]

These urban projects had an explicit environmental focus. The linchpins of the PUIs were the parks, schools, and libraries that functioned as architectural symbols of economic reinvestment and social transformation in the city's informal communities, while also helping to increase Medellín's heat-mitigating tree stock. Public buildings were designed with bioclimatic features such as cross ventilation and daylight capture, which help reduce energy consumption.[100] For example, the city's Moravia Park project reconstructed a neighborhood built around an old open garbage dump. Squatter communities living in the dump were moved to adjoining hazard-free areas, and a central hillock formed by garbage was recovered as a park using low-impact environmental technologies, afforestation, and sustainable drainage systems to restore the natural water cycle of the dump.[101] The connective and holistic approach that characterized Medellín's social urbanism made a major contribution to greening the city. In the Moravia project alone, for instance, over fifty thousand new trees were planted from forty-six species.[102] In addition, the city's Integrated Urban Projects dramatically increased people's security of tenure: home upgrading was yoked to land titling in Moravia, generating positive feedback between improved housing and more secure tenure, in a symbiotic relation that built community resiliency.

Perhaps the most celebrated instance of social urbanism, a design innovation that has led some commentators to see slums as models for sustainable living, is Medellín's cable car transit system.[103] Built in the city in 2004, the Metrocable gondola lift system links the relatively wealthy downtown to the slums clustered precariously on the steep hillsides of the urban periphery. Although the spectacular visual character of the cable car system has attracted tremendous attention, Metrocable is perhaps most remarkable for reintegrating the informal and formal city, with all the attendant implications for social justice. After all, the gondolas hoist people from the job- and amenity-filled downtown to the slums in about a quarter of an hour. This trip used to take many hours, with slum residents sitting in expensive buses that had to slowly navigate tight switchback roads up into the mountains.

Another important feature of the cable car system is its small environmental footprint. Researchers have concluded that the project reduced carbon emissions by over 17,000 metric tons per year as compared with diesel oil-burning buses that most people once used.[104] Along with related innovations such as open-air electric escalators, Bus Rapid Transit, and an emphasis on transit-oriented development in general, the cable car system typifies new connectivity schemes put in place by radical political movements that are linking the informal and the formal city, in ways that significantly improve the everyday experiences of slum dwellers. At the same time, the project enhances resiliency and reduces greenhouse gas emissions. For this reason, informal cities of the Global South have come to be seen as important laboratories for experimenting with low-cost— and even, in some cities, free—public transit, offering powerful examples of how low-carbon infrastructures can help reknit the urban fabric.

## The Right to Urban Mobility

Medellín's cable car system (which has been emulated in many other cities in Latin America) underlines that there can be no right to the city without a right to mobility. Indeed, the spatial disenfranchisement of urban dwellers is among the most palpable and onerous manifestations of urban inequality. To get a visceral sense of why this is true, we need only return briefly to the lives of slum dwellers in Lagos depicted in Chris Abani's *Graceland*. Abani's protagonist, Elvis, boards a bus to travel from his home to the city's swank hotel district, where he dances for money. The bus Elvis rides in is a rolling embodiment of the acts of bricolage that characterize subaltern urbanism, typifying the city's capacity to hybridize and refunction the detritus of Western modernity:

> The cab of the bus was imported from Britain, one of the Bedford series. The chassis of the body came from surplus Japanese army trucks, trashed after the Second World War. The body of the coach was built from scraps of broken cars and discarded roofing sheets, anything that could be beaten into shape or otherwise fashioned. The finished product, with two black stripes running down a canary body, looked like a roughly hammered yellow sardine tin. The buses had a full capacity of 49 sitting and 9 standing, but often held 60 and 20. People hung off the sides and out of the doors. Some even stood on the back bumpers and held onto the roof rack.[105]

These jam-packed, privately run vehicles (known in Lagos as *molue* buses) are not the stuff of fiction: they are the major means of communal mobility across not just Lagos but many of the cities of the Global South. They move large numbers of people around, offering the poor forms of flexibility in terms of routes and times that a city's deteriorating public transit system—if a system even exists—cannot provide. But there is a toll: run on a free-market basis, private transit vehicles often absorb a significant quantity of poor people's wages. And such minivans and buses are extremely hazardous. As Abani's novel shows, death is a regular part of life for urban commuters in Lagos and similar cities.

Walking is by far the most important way of getting around in the cities of the Global South, where, depending on the city, between 35 and 90 percent of trips are made on foot.[106] In cities such as Johannesburg and Mexico City, roughly half of residents are estimated to be underserved by existing modes of transit.[107] This means that they must endure obstacles including gruelingly long and often unsafe walks, long waits between badly connected services in inconvenient locations, and expensive trips in crowded and unsafe vehicles. Poor people in the cities of the Global South spend up to 35 percent of their income on transportation, but even middle-income people in these countries are forced to endure excruciatingly long commutes from peripheral suburbs.

Adding to this challenging picture, transportation patterns are changing at lightning speed in these cities. Globally, over sixty-six million cars and light trucks were sold in 2021, but these numbers are down from the pre-pandemic high of ninety-five million cars and light trucks in 2018. The total number of vehicles on the world's roads is expected to double from these pre-pandemic highs by 2030, and to triple by 2050.[108] City governments have often responded to urban growth by supporting car-based development through building new road capacity, locking in urban development patterns that impose fossil fuel–based private vehicle use, even when city dwellers would prefer alternatives.

This is another example of the kind of malign path dependency that I discussed in the previous chapter in relation to agriculture. In this instance, the automobile-based maldevelopment that characterizes US cities like Los Angeles is becoming a global norm. The massive growth in private vehicles is not only expensive and inconvenient, it also actively destroys the urban fabric, both by undermining the forms of street-based conviviality that are an essential feature of vibrant cities and by worsening dangerous urban environmental phenomena such as air pollution and the urban heat island effect.[109] And, of course, the global spread of the automobile has dire implications for the climate crisis: as energy and transportation democracy activists Sean Sweeney and John Treat report, transport emissions are projected to grow two to four times faster than overall emissions in Global South countries.[110] This means more congestion, but also, ultimately, more human and environmental devastation as climate breakdown wreaks havoc on the world's cities.

Buildings and transportation are key environmental features of cities, but the two are distinct in important ways. While, as we have seen, buildings can be and are autonomously constructed by communities, at first glance mass transit seems far more tethered to the economic and political power wielded by the state. After all, specific communities or even alliances of cooperatives cannot construct citywide subway or rapid-transit bus systems. There certainly is a role for the state, both on a municipal and a national scale, in rolling out carbon-free mass transit in cities. As the authors of the report *From Mobility to Access for All* argue, cities need to stop supporting private vehicles and find ways to manage demand for such modes of transportation.[111] "Private car and motorcycle use," they state, "is systematically underpriced, which translates into a de facto subsidy by all taxpayers for a mode of transport used by a minority of residents."[112]

Cities can reverse these harmful subsidies by discouraging private vehicle use in dense city cores, using policies such as adequate pricing of car use and parking in these areas. But in tandem with efforts to reverse the prodigious growth of private vehicles, cities need to build out an ecosystem of low-carbon, integrated transport services so that people have alternatives to private vehicles.[113] Such public transit should be convenient and affordable—or, even better, free. As we have seen, the creation of such transit systems has been a key element of social urbanism during the last two decades. Innovations developed in cities like Medellín and Curitiba in Brazil include not just cable car systems but also Bus Rapid Transit systems or enhanced bus corridors with priority lanes.[114]

But a significant amount of mass transit in the cities of the Global South is informally organized and unfolds below the radar of urban planners and the state. If urban transit systems are going to be integrated, if transit is to include the billions of slum dwellers across the Global South, it must include and upgrade forms of transportation such as the *molue* bus that Elvis takes in Lagos.

Decarbonizing cities means not just acknowledging collective, community-led modes of informal transit but also supporting demands for democratic, egalitarian streets where everyone's personal safety is guaranteed. In Quito, Ecuador, for example, the women's collective Carishina en Bici (a Quechua phrase that translates as "bad housewives on bikes") is part of a community-led ecosystem of informal transportation that includes bicycles, microbuses, minivans, sedans, and motorcycles.[115] As urban studies scholar Julie Gamble documents, Carishina en Bici brings together women of all ages, ethnicities, and backgrounds to organize for people's right to move around the city streets free of sexual harassment and fear. While the organization is explicit in its demands for decarbonized transportation, it is not explicitly engaged in climate politics, as some of its allies are. It is, however, part of a network of popular low-carbon transit options that are established beyond the realm of the state. These communal options play a critical role for communities not included in the city's large-scale, sustainable transport planning, which all too often is centered on affluent central city areas.

Currently, urban planners and city authorities tend to ignore informal transit networks, or, worse, in a replication of their attitudes toward informal settlements, they tend to represent them as illegal and try to eradicate them. To reverse these misguided and futile policies, cities should find ways to electrify the services of informal transit companies, and should also develop ways to integrate them into formal sustainable transit options. The efforts of autonomous collectives like Carishina en Bici to educate city planners about what a fearless, feminist, democratic, and decarbonized city should look like are exemplary of broader struggles for an empowering urban future. They also return us to the figure of Marie-Sophie Laborieux from the novel *Texaco*. Laborieux's struggle to protect her community involves telling tales that reframe urban history, rendering the invisible visible, the subaltern heroic, and the impossible possible. We need many more such tales.

## How to Build Sustainable and Just Cities for All

It is worth returning, by way of conclusion, to Mike Davis's pivotal question: Who will build the ark? As he puts it, "I appeal to the paradox that the single most important cause of global warming—the urbanization of humanity—is also potentially the principal solution to the problem of human survival in the later twenty-first century."[116] Davis's essay surveys the contemporary urban condition looking for the agents of revolutionary transformation, the contemporary Noahs who will build an ark to carry humanity through the coming deluge. He concludes that "a new Ark will have to be constructed out of the materials that a desperate humanity finds at hand in insurgent communities, pirate technologies, bootlegged media, rebel science, and forgotten utopias."[117]

We have seen over the course of this chapter that Davis's hoped-for ark is in fact already being built by urban climate insurgents in the Global South, using whatever materials they have at hand in acts of ingenious bricolage. But these feats of autonomous construction are carried out in the face of a state that often tries to evict and expunge squatter communities. As the climate crisis escalates remorselessly, the enormous challenges these communities confront, from lack of clean water to intensifying urban heat, are only going to grow.

The cities that are best situated to weather this strengthening storm are those where strong social movements build autonomous power, to the point where they can force elected leaders to place the resources of the state at the disposal of urban climate insurgents. We have seen how such movements can transform cities when they are able to mobilize the technical know-how of architects and urban planners alongside their visions of a more just and sustainable city. In this way, Global South cities can become a laboratory for prodigiously creative solutions to the urban climate crisis. But these solutions will only germinate and spread in the fertile soil of cities and countries where movements of urban climate insurgency can pressure the state to cease and reverse anti-poor policies.

This means that a lot will depend on whether popular movements can build their power and expand the appeal and traction of radical ideas in the face of the worsening climate emergency. Should they fail to do so, the crisis will strengthen reactionary political and economic forces that are spreading climate change denialism and eco-fascism.[118] This question will play out within particular cities, countries, and on a global scale. As I acknowledge in the introduction to this book, radical movements' arguments for global climate reparations over the last several decades have borne scant fruit. The elites who control imperial countries may be aware that they and their ancestors have colonized the atmosphere, but they show few genuine signs of being willing to pay back these debts. What would be possible if they could be shifted, if the political calculus began to swing toward the payment of reparations to Global South nations? And how might such reparations play out on an urban scale?

The proposal for a million climate jobs, advanced by the Alternative Information and Development Centre (AIDC) in South Africa, offers a tantalizing blueprint for the wholesale ecological reconstruction of cities, one that scales up many of the initiatives discussed in this chapter.[119] The capitulation of the ruling African National Congress to neoliberal austerity policies in the 1990s and the party's shocking corruption in recent years has made it virtually impossible to translate this blueprint into transformative policies for the country's cities. Nevertheless, the AIDC proposals continue to be an inspiration to important segments of the country's Left, including, as we will see in the next chapter, some of its major labor unions. For this reason, the One Million Climate Jobs plan remains an impactful political document, both within the country and across the Global South.

Among the useful AIDC suggestions is the proposal that, with a combination of funding from the South African government and climate reparations, municipalities could map communities, informal settlements, and other infrastructure that is at risk from the climate crisis.[120] This, they note, would include generating an accurate picture of communities whose lack of access to water, electricity, and housing will be worsened by climate change. Such a centralized database could provide local communities with valuable knowledge about present and future environmental challenges.

The task would then be fairly straightforward in principle, if monumental in scale: cities would need to convert old buildings so they use much less energy, build new ones that use virtually no energy, and ensure that the built environment in general is able to cope with the growing consequences of climate change. Responses to this task will vary: particular cities must come up with their own solutions to their problems, while also appropriating and transforming the best responses developed in other places.

The ecological reconstruction of poor cities could, as the AIDC argues, generate millions of jobs in Global South nations, in the process generating hope and a sense of possibility. The new houses to be built by local workforces, for example, should be constructed with top-notch insulation, in-roof solar panels, solar water heaters, and rainwater collection. As the AIDC argues, "The equivalent of the 'passive houses' now built for some affluent Germans could become the birth-right of working-class South Africans."[121]

In urban transport, the AIDC argues that transit not only needs to be cheaper than private vehicles, but that it needs to be clean, safe, and even enjoyable.[122] The *molue* buses that Elvis rides in Abani's novel give us a sense of how public transit can be a space of important and amusing social interaction. If only riders did not have to wager their lives when boarding such buses. Moving around the city should offer citizens a form of corporeal experience and social interaction that they can look forward to. This means not only reversing the terminal underfunding of public transit by Global North–dominated entities like the World Bank, but also integrating transit into broader movements for democratic and fearless cities.

Last of all, the AIDC reminds us that the structural forces in the global economy that have generated planetary urbanization in the first place need to be tackled head-on. South Africa, the AIDC argues, is a nation built on the extraction of cheap energy and cheap labor.[123] In this regard, it is similar to the majority of other Global South nations. The dispossession of South Africa's Black majority produced cheap labor that was employed to extract minerals such as gold and diamonds for export abroad, but it left the country in a state of semi-industrialization and with intense inequality and poverty, meaning that most South Africans are too poor to buy many of the goods made by local factories. This savage regime of rural dispossession, mass impoverishment, and export-oriented capitalist accumulation enriched a small elite but generated massive displacement to South Africa's cities. As in other nations, this brutal arrangement is unsustainable in economic and ecological terms, and is fast unraveling.

The peoples of the Global South need a new path beyond this failed and oppressive model of extractivism. If the human condition is now an urban condition, how will we power the cities of the future? The next chapter surveys the devastating impact of fossil fuel extractivism across the Global South, and explores the alternative models of development that are being built by contemporary struggles for environmentalism from below. If fossil capitalism seems hell-bent on pushing the planet toward a terminal environmental crisis, people's movements of environmentalism from below are fighting for a way out of this deadly trajectory, and for a future grounded in revolutionary power.

# CHAPTER 3

# RECLAIMING THE ENERGY COMMONS

In 2001, the Soweto Electricity Crisis Committee (SECC) mobilized thousands of activists to fan out across South African townships with tools and connector cables, risking death by electrocution to reconnect working-class households to the grid.[1] Operation Khanyisa! (Zulu for "enlightenment") was launched after the country's Electricity Supply Commission, still known by its Afrikaans acronym, Eskom, cut off electricity to people who were unable to pay their utility bills in full. The power cuts had left more than three thousand families huddled in the cold and darkness. As SECC co-founder Trevor Ngwane explained to me, Operation Khanyisa! was a grassroots community response to a social crisis sparked by Eskom.[2]

After the country's first democratic elections in 1994, Ngwane and other community activists had negotiated fixed payment rates with Eskom. In 2000, however, as part of a broader shift toward neoliberal modes of governance by the ruling African National Congress (ANC), Eskom abandoned fixed-rate charges for electricity and began charging fees based on kilowatt hours consumed.[3] In tandem with this shift, the price of electricity shot up as the ANC reintroduced the post-apartheid South African economy to competitive global markets. Township residents, including many elderly people on pensions, could not afford the new rates and began to fall into arrears. Eskom reacted by shutting off power to those who couldn't keep up with payments, justifying such punitive action by alleging that township residents were deliberately refusing to pay their bills as a form of boycott, a hangover tactic from the anti-apartheid era, or what Eskom officials called a "culture of non-payment."[4]

After forming the SECC in response to these cutoffs, Ngwane and his comrades tried negotiating with Eskom for clemency for individual households. Ngwane told me that he soon realized this was futile since the problem was structural: township residents were simply too poor and electricity was too expensive. Every time Eskom granted debt forgiveness to one household, it cut off power to at least two others. The result was mass debt and widespread disconnection from power. Meanwhile, Eskom continued to provide discounted electricity to big business, including the country's coal, diamond, and gold mines.

The impact of the termination of electricity, one of the fundamental elements of urban modernity, is both material and psychological. As a recent report on the social impact of power outages in South Africa documented, power cuts targeting the country's poorest and most vulnerable communities mean widespread disruption of social reproduction.[5] People are forced into a stressed, adaptive mode of thinking that eats up precious time while also sabotaging their capacity to think long-term. As one township resident put it, power shutoffs mean that "we are dying from the inside slowly, even if we did not go to the forest, we are like dead people inside."[6] Electricity cutoffs are a kind of slow death, a targeting of those rendered structurally marginal to the economic and social order, denying them access to the material and social infrastructures essential to life.

In response to these oppressive conditions, the SECC insisted that access to a basic amount of electricity was a human right. Community activists saw their work as a form of mutual aid predicated on "commoning" electricity.[7] Ngwane and other SECC activists argued for the decommodification of electricity, insisting that the ANC follow through on the promises of free electricity it had made during the 2000 elections. Other demands included a reversal of the government's moves toward the privatization of Eskom, which activists felt had led to rate increases; the addition of special provisions for vulnerable groups like pensioners and HIV-positive people; and, lastly, the provision of solar power by Eskom to all township households.[8]

In making these demands, community activists were responding not only to immediate community needs but to the South African government's own declared ambitions: in a 1998 document titled *White Paper on Energy*, the government had promised to "promote access to affordable energy services for disadvantaged households, small businesses, small farms and community services."[9] The ANC at the time of the report saw universal access to affordable energy as a matter of environmental and social justice. Poor households in South Africa had long depended on fuelwood and coal for domestic use, generating indoor air pollution with serious health effects. Officials saw that access to electricity would ameliorate the deadly effects of air pollution; however, beyond that, it was key to transforming the inequality generated by apartheid. Affordable energy was going to be made available not just to the country's poorest but to all citizens as part of the nations' post-apartheid economic and social reconstruction.

Yet despite the government's promises, Eskom and the ANC refused to listen to the demands of groups like the SECC. In response, the SECC launched Operation Khanyisa! to reconnect people to the grid and pressure the government to reverse neoliberal policy. The campaign was an initial salvo in what was to become a much broader campaign of protests against the ANC's failure to deliver on its promises to provide basic services such as water, education, housing, and electricity to its constituents.[10] Sometimes called "service delivery" protests, these rebellions of the poor have become remarkably widespread: from 2014 to 2017, for example, there were 1,500 community protests per year in South Africa, an average of four per day.[11] These protests are often nonviolent but can also include the revival of militant tactics from the anti-apartheid era, including burning tires, barricading roads, and vandalizing property.

Service delivery protests reflect the politicization of infrastructure in the post-apartheid state. They constitute an angry popular response to what many perceive as a revolution betrayed. In South Africa, the state's failure to provide basic services to township residents is a palpable sign that the ANC has reneged on the emancipatory and egalitarian promises on which it was founded.[12] The result is a crisis that is unfolding on a political scale, as support for the ANC fragments, competing parties such as the Economic Freedom Fighters arise, and activists turn to community mobilization against a state that is again seen as an enemy of the people rather than an embodiment of their thirst for liberation.[13]

Equally importantly, revolutionary betrayal plays out on an emotional and moral level. Denial of access to electricity and other basic elements of urban infrastructure catalyzes what could be called "broken hope," a state of anger and moral outrage that results when a liberation movement gains power but, instead of fulfilling its promises to the people, lets infrastructure rot and collapse, all the while pilfering from the public to enrich a small gang of cronies. Broken hope is what succeeds the cruel optimism stoked by national liberation movements that choose complicity with a neoliberal capitalist world order dedicated to extreme extraction.[14]

Broken hope might perhaps be seen as a prototypical postcolonial condition, since so many movements for national liberation have transformed into authoritarian oligarchies. In India, it is evident in the history of the building of hundreds of big dams in places like the Narmada Valley, submerging huge tracts of land and displacing hundreds of thousands of people without delivering the abundant water and energy promised by engineers and political leaders.[15] The corruption and injustice that characterize such energy-generating megaprojects are evident in many other Global South nations, from the Democratic Republic of Congo to Colombia. Such betrayal by elites is not new. Indeed, anti-colonial writers such as Frantz Fanon warned of the pitfalls facing postcolonial countries, writing that instead of supporting popular engagement in a process leading to economic redistribution and participatory democratization, nationalist leaders faced strong temptations to enrich themselves while seeking to pacify the people and silence dissent through "baton charges and prisons."[16] Fanon decried the tendency of the postcolonial bourgeoisie to betray their compatriots, working with imperial countries like the US to continue the plunder of their countries despite formal freedom from colonialism. His words, written during the high tide of decolonial movements, proved all too prescient.

Yet, while broken hope may be a long-standing syndrome besetting postcolonial nations, the betrayal of revolutionary aspirations is particularly momentous in the present. As a result of global warming, droughts and famines are spreading across the globe. Some parts of the world are even becoming too hot and humid for human survival.[17] Lack of access to basic infrastructure can be an immediate matter of life and death at a moment of intersecting environmental crises. Denial of water and electricity is effectively a mass death sentence, a form of necropolitics already affecting the poor in cities across the Global South.[18]

What's more, struggles over access to electricity are interwoven with fights against fossil capitalism whose stakes are nothing short of the planet's future. Environmental scientists have made it quite clear that most fossil fuel reserves must be left in the ground if a truly cataclysmic environmental crisis is to be avoided.[19] Nonetheless, the world is *not* transitioning away from fossil fuels. Instead, while growth in renewable energy has occurred between 1990 and 2015, it has been matched, and in many places outstripped, by growth in extraction and consumption of fossil fuels. Although renewables constituted 80 percent of all new electricity capacity added in 2020, for example, the share of renewables in total final energy consumption between 2010 and 2019 only grew by 2.7 percent.[20] Indeed, the dirtiest of fossil fuels, coal, contributed more than twice as much to global total primary energy supply as all renewable energy sources combined.[21] In many Global South nations in particular, coal is playing an increasingly decisive role in the energy mix, notwithstanding official rhetoric about energy transition.[22] The implications are stark because these societies are undergoing transitions from predominantly biomass-based to fossil fuel–based systems, locking in forms of fossil development at the precise time when the world needs to cease burning fossil fuels.

This chapter explores the complex stakes of the energy transition in Global South nations. The fight for energy sovereignty has been foundational to anti-colonial nationalism. There is a long history of imperial efforts to disrupt and pulverize movements fighting for control of fossil energy resources in Global South nations.[23] Wealthy nations' refusal to cease their colonization and pollution of the atmospheric commons by failing to slash carbon emissions is merely another round in these struggles. This fight is certainly not over, not only because of the intensification of fossil fuel extraction and exploitation that I alluded to above, but also because the transition to renewable energy is a complex proposition. The media in wealthy nations is filled with celebrations of the diminishing cost of renewables, a development which is seen as driving an inevitable, market-led transition toward clean energy. But such celebrations of green capitalist energy transition ignore the growing contradictions of energy systems around the world. Developing countries are following the model established by richer nations, where public funds are used to subsidize the installation of renewable energy generation on electric grids maintained by public utilities. As renewables grow cheaper, private investors see less opportunity for profits, and installation rates of renewables plummet. With the market for renewables currently in crisis in wealthy regions such as the European Union, this is not a model for successful or just energy transition that Global South nations should embrace. The current crisis of South Africa's public energy utility, Eskom, offers a dramatic example of this failing model, as we'll explore in greater detail below.

Ruling elites in poorer nations tend to argue that the continued extraction of coal and the privatization of the electricity system are necessary to meet the "needs" of the people, and of the poor in particular. The struggle for universal electrification overlaps to a remarkable extent with many postcolonial nations' sense of their place on the world stage. Leaders of postcolonial nations see electrification as a vital step in the transition from a subordinate position as exporter of primary commodities to a competitive position in the capitalist world-system. They consequently tend to assert a right to power, in both senses of the term. Energy sovereignty is equated with a broader quest for national autonomy and development.

Even governments with a strong orientation around climate justice have adopted policies of "social" extractivism, through which the profits of natural gas and mineral exports are used to improve public infrastructure and alleviate poverty through redistributive policies.[24] Such policies have won significant popular support. When challenged about the environmental effects, leaders of Global South nations often point out that the core imperial nations developed by burning as much fossil fuels as they could get their hands on, and they still refuse to slash their carbon emissions. What right do nations made wealthy by fossil capitalism have to reprove nations of the Global South for developing their own fossil energy resources?

These arguments are self-serving nonsense. When elites set their nations on this path, they are ushering their more vulnerable fellow citizens—peasants, fisherfolk, Indigenous people, women, and children—to the brink of the abyss. The embrace of fossil capitalism in the name of development is a grand and ultimate betrayal. As I will show in the stories that follow, the poisonous promise of fossil fuel-based energy sovereignty leads directly to oblivion. Fossil developmentalism is an oxymoron, a political program for mass death.

Will the masses in the Global South go along with fossil developmentalism or will they fight for alternative forms of power? This chapter explores anti-extractivist campaigns in South Africa and India that constitute what I term *Global Blockadia*. Popularized by Naomi Klein in her book *This Changes Everything*, *Blockadia* refers to movements against fossil capitalist infrastructure that have helped to normalize direct action in land defense over the last decade.[25] For Klein, Blockadia is "a roving transnational conflict zone" in which people embrace radical forms of direct action, in contrast with established environmental organizations' conventional reliance on awareness-raising and advocacy efforts with lawmakers. As we have seen in earlier chapters, environmental movements in the Global South have long deployed the kinds of direct-action tactics that have only recently been embraced by Blockadia in the rich countries. Yet anti-extractivist environmentalism from below is nothing new in the Global South.

In addition, Blockadia is not spread evenly across the globe. A comprehensive study of global conflicts related to both fossil fuels and low-carbon energy projects like big dams concluded that these projects disproportionately impact peripheral countries and regions, including rural areas. These energy projects also most strongly affect Indigenous peoples, minorities, and those who depend on nature for their livelihoods.[26] Although Indigenous people constitute only 3 percent of the global population, for example, they were impacted in over half of the cases documented by the Global Atlas of Environmental Justice project. And they are more likely to engage in direct action to shut down such extractivist projects.

Anti-extraction activists in various countries are increasingly building ties with the global movement for climate justice, but it should be recognized that such solidarities must be built across significant geographical, economic, and cultural differences. Solidarity in Global Blockadia derives from the active political construction of relations between diverse places, activists, and groups—not solely, as is sometimes assumed, from cultural similarity or likeness.[27] Solidarity is dynamic and inventive. It is elaborated across uneven power relations and geographies, and in the process generates new political relations and spaces.

Perhaps most importantly, Global Blockadia is succeeding: over a quarter of fossil fuel and low-carbon energy projects documented by the

Global Atlas of Environmental Justice were canceled, suspended, or delayed when met with social resistance.[28] But these victories come at a steep price. Violence against protesters is rife and hydra-headed. It is most commonly used against people and communities resisting coal extraction and oil pipelines, as well as hydropower and biomass projects.[29] Ten percent of all the cases analyzed by Environmental Justice Atlas project scholars involved assassinations of activists.

Global Blockadia is a key alternative to neo-extractivism. An ideology espoused by elites in Global South countries, neo-extractivism asserts that continued extraction and exploitation of minerals and fossil fuels will benefit the people and offer a viable pathway for national development. Given increasing demand for energy in much of the Global South, Blockadia's rejection of fossil infrastructure needs to be accompanied by viable proposals for a just transition to renewable power. In many Global South countries, terms like *just transition* are regarded with significant skepticism given sky-high rates of unemployment and the tendency for the renewables sector to be controlled by private, for-profit corporations rather than democratically controlled organizations dedicated to the public good.

But energy transition need not involve a new round of the kind of hollow system of governance that historian Timothy Mitchell calls "carbon democracy."[30] Down with the oiligarchy! Another world based on energy democracy is possible. Publicly owned and democratically managed renewables must be part of a just transition. As the cases of South Africa and India demonstrate, diverging models for energy transition always open out onto fundamentally political questions about governance and democracy. Global South struggles against extractivism demonstrate that the fight for clean energy is just as much a struggle for popular power as it is an effort to abolish fossil capitalism.

## "Mining Will Be Our Death": Coal and Dispossession in India

Early in the morning one day in the spring of 2022, people living in villages surrounding the Hasdeo forest in the central Indian state of Chhattisgarh were woken up by the sound of heavy tree-cutting equipment tearing through the woods. The state government in Chhattisgarh had approved the digging of a coal mine in the area, even though the national Ministry of Forest and Environment and the Ministry of Coal had declared the area a no-go zone in 2010 because of its immense biodiversity.[31] The local people of this densely forested region are Adivasis, Indigenous people who are often referred to as "tribals" in India. By the time the villagers got to the spot where trees were being felled, as many as three hundred had already been cut down. Police officers accompanied the men who came to raze the forest. Even when they were shown approvals for the logging, the Adivasi villagers continued to protest by putting their bodies between the logging equipment and the trees. This nonviolent tactic eventually drove the loggers and police from the area.

Since then, hundreds of Adivasi residents of the region have been camping out, day and night, to protect their forest. Alok Shukla of Chhattisgarh Bachao Andolan, an activist organization working against mining in the area, explained, "Our estimate is that not only will hundreds of people be displaced due to these two new coal mines, as many as 4.5 lakh [hundred thousand] trees will also be cut. We are fighting against this horrific tragedy on the ground and in the courts."[32]

Protests to protect the Hasdeo forest have been going on for over a decade. As Bhual Singh, a villager likely to be displaced by extraction projects in the forest, lamented during an earlier protest, "Mining will be our death."[33] As many as five billion tons of coal lie beneath Hasdeo, a 420,000-acre forest that is home to ancient trees, hundreds of elephants, and much other wildlife. Some of India's biggest mining conglomerates, including the massive Adani group, have been vying to gain access to the coal. Governments in Chhattisgarh and the neighboring state of Rajasthan, led by the Congress Party, opposed mining projects for a number of years, accusing the opposition BJP party of corruption and of favoring Adani. But in 2022 the Congress-led state government in Chhattisgarh gave the go-ahead for the Parsa coal mine after receiving a green signal from the central government the previous year. Neither of the country's dominant political parties, in other words, could be relied on to support Adivasi efforts to protect their land and the forests that are of critical cultural and material importance to Adivasi people.

The Indian government's plans to double coal production will require selling off rights to vast areas of forest where Adivasi people live. On June 18, 2020, the Indian Ministry of Coal announced a plan to "Unleash Coal." In an unctuous press release, the ministry praised the "visionary and decisive leadership" of Prime Minister Narendra Modi and stated that the auction of the rights to develop new coal mines would be "a historic day when [the] Indian coal sector would break free of the shackles of restrictions to charter new growth."[34] The press release described the plan, which provided for the auctioning of rights to develop forty-one coal mines, many in biodiversity-rich areas of India, as an effort to "make the country Atma Nirbhar (self-reliant) in coal mining."[35] Ironically, what made this announcement truly historic was that it was the first time that private firms would be allowed to mine coal for commercial purposes without any end-use restrictions, meaning that both Indian and foreign mining corporations could sell the coal they dug up for export as well as for domestic use, despite the nationalistic language of self-reliance used to justify the auction scheme.[36]

India's plans to massively expand coal extraction could not come at a worse time when viewed through the lens of the climate emergency. According to climate scientists, 90 percent of existing coal reserves must remain unextracted to keep the world within a 1.5° carbon budget.[37] The dirtiest of fossil fuels, coal is a particularly potent and lethal driver of climate change, but 60 percent of oil and fossil methane gas also need to be kept in the ground to avoid climate cataclysm. UN Secretary-General António Guterres consequently made an urgent appeal in 2021 calling on wealthy nations to shut down all coal-fired power plants by 2030.[38] All other countries—including historically colonized and poorer nations like India—must cease burning coal by 2040, he told members of the "Powering Past Coal" Alliance.

The world is going in precisely the opposite direction. As nations rebounded from the pandemic, carbon emissions rose to record levels.[39] While it is true that renewables like wind and solar power have been growing rapidly, such growth has been outstripped by the growth of the total primary energy supply (TPES) derived from fossil fuels.[40] In fact, as noted above, the expansion of coal's contribution to TPES was twice that of all renewable energy sources combined.[41] While coal production has declined and consumption is being phased out in some parts of the Global North, new coal energyscapes are emerging in nations across the Global South. These emerging coal geographies are often driven by a combination of repressive governments and powerful corporations, a toxic alliance that results in intense environmental injustices.[42]

The failure of genuine energy transition is particularly apparent in India. Despite a much-publicized renewable energy program that Prime Minister Narendra Modi has promised will provide 50 percent of the country's energy by 2030,[43] in 2017 India was the world's second largest producer, consumer, and importer of coal.[44] Between 1994 and 2015, the national total of coal extraction doubled, from approximately 250 million tons per year to 500 million, increasing coal's contribution to TPES from one-third to one-half.[45] Coal combustion generated 72 percent of India's electricity and was responsible for 65 percent of its carbon dioxide emissions. India's coal industry is to a significant extent controlled by the government, which is continuing to expand investments in this, the dirtiest of fossil fuels, despite its rhetoric about the shift to renewable power: in 2020, the Indian government announced a $6.5 billion investment in the coal sector, locking in coal extraction to the tune of an estimated billion tons by 2023–24.[46]

India's decisions about energy pathways are hugely consequential for the country itself, but also for the world. With a population of 1.3 billion, India's energy needs are set to rise more than any other nation on Earth over the next twenty years, according to the International Energy Agency (IEA).[47] As IEA head Fatih Birol puts it, "All roads to successful global clean energy transitions go through India."[48] The country's coal extraction has made it an antagonist of climate action. At the 2021 UN Climate Summit in Glasgow, for example, India joined China in a last-minute demand to water down language in the Climate Pact calling for a "phase-out" of coal.[49] Climate activists such as Brandon Wu of ActionAid rightly challenged the hypocrisy of wealthy nations who castigated India and China for their obstructionism even as they refused to include language about shutting down oil and fossil gas.[50] But the hypocrisy of the core imperial nation should not make activists forget that a rapid transition away from fossil fuels needs to happen everywhere, including in poorer nations.

India's anti-imperialist stance in its proclamations of energy sovereignty, reflected in the Coal Ministry's messaging about national self-reliance, has a material basis: since the country lacks significant oil and gas reserves, it spends more than $100 billion each year to import these fuels. As Indian officials point out, core imperial nations like the US developed by exploiting fossil fuels, essentially colonizing the atmosphere by filling it with the pollution produced by their fossil-based economic development. India now needs "carbon space" to meet its development goals, in the words of Sandeep Pai, researcher for the Global Just Transition Network.[51]

But coal-based development is not improving the lives of the majority of India's inhabitants. India is the world's fifth most vulnerable country with respect to climate extremes.[52] Of course this vulnerability is not distributed evenly: the 60 percent of the nation who live below the World Bank's median poverty line are particularly susceptible to climate calamity.[53] India had the greatest number of deaths in the world in 2018 due to extreme weather events caused by climate change, from cyclones to floods and heavy rain–induced landslides.[54] The country's economic losses that year as a result of climate change were the highest in the world. According to analysis by India's Council on Energy, Environment and Water, over 75 percent of the country's districts—home to 638 million people—are climate-related extreme event hotspots. The frequency of flood-associated events surged by over twenty times in the period from 1970 to 2019.[55] Moreover, India is already being hit hard by the results of fossil-based economic growth: the terrible air quality in the nation's cities, a result of burning highly polluting coal, kills more than one million Indians each year, and reduces the life expectancy of the average Indian by a full five years.[56]

Corporate priorities and profits have come to define India's "development." As political analyst Shankar Gopalakrishnan has argued, "Large projects involving speculative or extractive use of natural resources were equated with 'development' (through such notions as 'employment generation'—even when it was clear that no net employment was being generated—or 'infrastructure provision')."[57] In line with this increasingly predatory mode of capitalism, the Hindu-fundamentalist BJP government has adopted policies of extreme extractivism, reflected in but also going far beyond the plan to "unleash coal." Under Modi, the government has promised faster clearances, faster permissions, and "ease of business" for extractive projects, and has done its best to undermine, sabotage, and nullify laws protecting the country's environment, all in the name of development.[58] Those who inhabit the lands under which coal is buried, India's Indigenous or Adivasi people, are at the sharp end of this extractivist surge.

But mining and other projects on Adivasi land legally require the consent of those communities. This requirement is a result of a successful struggle for landmark legislation, the Scheduled Tribes and Other Traditional Forest Dwellers (Recognition of Forest Rights) Act, commonly referred to as the Forest Rights Act (FRA). The FRA became law in India in 2006 following the political defeat of the BJP-led National Democratic Alliance in national elections two years earlier. According to Shankar Gopalakrishnan, the surprise defeat of the BJP-led alliance and its "India Shining" rhetoric constituted a significant ideological setback for neoliberal ideology in India.[59] Although this electoral defeat by no means overthrew the power of the corporate sector and its BJP allies, it did create enough political space for an alliance of Adivasi and forest dwellers' movements to garner broader support for a law that would curtail corporate plunder while establishing more democratic and collective control over the forest commons. The broader implication of the FRA was to challenge dominant conceptions of both the "environment" and "development," which tend to be seen as antithetical, with the latter destroying the former, and protection of the former impinging on the latter. The fight for the FRA led to a counter-articulation of both these terms by a broad coalition of social movements anchored by Adivasi activists.

As we will see in greater detail in the next chapter, India's forests are managed by the Indian Forest Service, an elite bureaucracy that dates back to the days of British colonialism in the mid-nineteenth century. The forest bureaucracy largely continues to adhere to a colonial model of forest management, in which its mission is to protect forests against irresponsible encroachers and destructive local communities. This official version of environmentalism sees the forest through the eyes of the state, to invoke James C. Scott's classic study of how the modern nation-state seeks to rule by employing standardizing forms of governance.[60] Exclusion and expropriation are central to environmentalism, in the eyes of this state bureaucracy. Environmentalism and dispossession are consequently not contradictory—indeed, they are in fact mutually constitutive and foundational to forest management in India.

In the run-up to passage of the FRA, these dominant views came under intense criticism from Adivasi and forest dwellers' movements, who pushed for specific provisions on community rights, collective management of forests, and the empowerment of the *gram sabha* (the village assembly) in determining rights. Although some of these demands were not won, the version of the FRA that made it into law nonetheless constituted a decisive rejection of dominant forest governance philosophy and practice. For the first time in the country's history, the state statutorily recognized forest dwellers' right to protect and manage forests. In addition, rights to the forest were to be determined through a process that begins in villages. This shift was immensely consequential. As Gopalakrishnan puts it,

> The political import of the FRA goes well beyond its text; it offered the possibility of a new system of forest management, and implicitly a new system of resource management, that went beyond the prior dominant discourses of both "environment" and "development." . . . Contained within it are the seeds of a genuinely different, collective, and democratic model of the use and conservation of nature, and of the livelihoods of people.[61]

The political window that facilitated passage of the FRA did not remain open for long. The forest bureaucracy sought to stymie the FRA from the beginning, often ignoring the law's requirement that village councils give consent before initiation of large-scale projects in the forests. Then, in 2014, the BJP-led National Democratic Alliance won national elections, sparking a return to full-scale resource speculation and extraction, carried out under the mantle of "ease of business" and "development" of the nation.

The extraction of coal and other minerals in India is made possible by legal traditions that grant the state sovereignty over all the resources that lie beneath the surface of the land, no matter who or what may occupy that land itself. These laws trump the Forest Rights Act. India's Coal Bearing Areas (Acquisition and Development) Act, or CBA Act, grants the government the right to acquire land after it determines that coal can be extracted from a certain area. There is no requirement to consult affected communities, or to seek the free, prior, and informed consent of Indigenous peoples, as required by international law.[62] The government's intention to acquire and develop land is listed in the official government gazette; anyone who objects to such acquisition has thirty days from the time of the listing to file an objection with the Ministry of Coal, which considers any such objections and then makes a recommendation to the central government. The state then decides whether to acquire the land, essentially by eminent domain. There is no requirement for authorities to pay compensation before taking possession of land. No human rights impact assessments are conducted prior to land acquisition, which means there is no protection for communities from forced eviction.

The CBA Act dates to the early years of India's independence, but, as the historians Matthew Shutzer and Arpitha Kodiveri have argued, it builds on the declaration of subterranean sovereignty promulgated by the British East India Company during the colonial era.[63] The postcolonial Indian state's right to summarily expropriate land for extraction, in other words, originates in colonial violence and expropriation. These acts of enclosure of an underground energy commons are justified in the name of national development.

The implications of this vision of development are stark in the tribal lands of central India. It gives implicit license to state authorities to proceed with development by any means necessary. The Chhattisgarh-based human rights lawyer Shalini Gera told me that it is quite common for government officials to forge the permission documents they are supposed to obtain from village councils.[64] Gera also said that it is normal practice for local officials to force village councils to meet while police armed with guns look on, intimidating them into agreeing to extractive projects. Her allegations are corroborated by Amnesty International and India's own Ministry of Tribal Affairs.[65]

But intimidation can take much more brutal forms. For example, on International Women's Day in 2021, hundreds of women gathered in a village in the Bastar district of Chhattisgarh to commemorate the death of two Adivasi women, victims of sexual assault by state security forces.[66] The mobilization challenged state security forces for their violence toward these two women in particular, but also for arrests, extrajudicial killings, and sexual assaults that have become daily realities for many Adivasi people in India's coal belt states of Chhattisgarh, Jharkhand, and Odisha. The police reacted to this protest by swooping down on the home of Hidme Markam, one of the protest's leaders, and arresting her. At the time of this writing, Markam has been held for over a year under India's Unlawful Activities Prevention Act (UAPA), a vague and virtually boundless anti-terrorism law that is being used to silence anyone who speaks out against the government.[67] The police accused Markam of being "an absconding Maoist insurgent."[68]

Such anti-terrorist charges are often leveled at activists who challenge state extraction policies. Shalini Gera told me that police made a similar accusation against a young Adivasi man named Badri Gawde in the area where she works. Gawde was imprisoned for years. Gera explained that police use of anti-terror legislation, as well as outright cold-blooded murder, chills efforts to assert community rights over the forests that Adivasis have inhabited for generations. The violence is systematic: the Asian Centre for Human Rights has documented thousands of cases of extrajudicial killings by Indian police.[69]

The state anti-terrorism campaign that targeted Adivasi activists like Hidme Markam and Badri Gawde in India's forests closely tracks the global war on terror. India's clampdown on activists uses the same broad language and facilitates the same draconian curtailments of legal and human rights employed by the US during that war. This similarity shows how nationalist projects of resource extraction can mobilize the US empire's anti-terror discourse for their own ends. In India, the principal overt target of anti-terror discourse and counterinsurgency violence is the underground Communist Party of India (Maoist), who are popularly known as Naxalites, after the village of Naxalbari in West Bengal where an armed peasant uprising took place in 1967. According to the anthropologist Alpa Shah, the Maoists first moved into the Adivasi-dominated forests and hills of Chhattisgarh and Jharkhand in the period from the early 1980s to the early 1990s.[70] Shah argues that the Maoists arrived in the forests from the plains of Andhra Pradesh, driven by the advantages of mountainous terrain for guerrilla struggle.

But while it was certainly easier for the Maoists to avoid capture in the dense forests of what came to be known as India's "red corridor," Naxalite leaders also found willing revolutionary subjects in the country's central states, which are dominated by Adivasi communities from whom lands had been stripped by the colonial and then postcolonial forest bureaucracy. As Shah explains, the Maoists' first move once they arrived in the forests was to burn the forest bureaucrats' jeeps and lodges, and to chase away the outside contractors whom forest officers collaborated with to remove truckloads of illegally felled timber. The Maoists thus effectively banished the main face of the oppressive nation-state, a force that had been targeting Adivasi communities for centuries.[71]

With the forest bureaucracy gone, Adivasi communities were once again able to access what have been classified as "reserved forests," opening them for grazing and the collection of firewood and bamboo, and allowing village committees to set their own rules for the use of forest products. In other words, the Maoists were not only able to liberate Adivasi communities from oppressive forms of state power that had excluded them from their communal lands since the mid-nineteenth century; they were also able to facilitate the reintroduction of direct democratic control over the forest commons. This transformation in the forests of India's central states effectively anticipated the democratic and anti-capitalist goals of the Forest Rights Act in the following decade.

Driven by a determination to extract the rich mineral resources lying beneath India's Adivasi-dominated forests, and aided by the anti-terrorist rhetoric of the global war on terror, the Indian state began to reassert its control over the red corridor beginning in 2005. The state's first move was to support a so-called people's movement named the Salwa Judum, which translates as "purification hunt" in the local Gondi language.[72] Salwa Judum vigilantes, accompanied by the police, marauded through villages in districts like Bastar, burning, looting, raping, and killing. The vigilantes also forcibly removed villagers to government-controlled concentration camps, in a replication of the forms of counterinsurgency practiced by occupying imperial troops in places like Kenya and Vietnam.[73] By 2009, the Judum had morphed into an openly state-sanctioned paramilitary force known as Operation Green Hunt.

As writers such as Nandini Sundar and Arundhati Roy have documented, the oppression of Adivasi communities, whether or not they were linked to the Maoists, metabolized over the subsequent decade, with anti-terrorist rhetoric providing a cover for the baleful alliance of Hindutva and corporate capitalism that lies at the heart of extractivism in India.[74] So virulent has the campaign against the Maoists become that in 2018 a group of prominent human rights activists residing in India's principal cities were arrested by police, who ransacked their homes and possessions while alleging that they were "urban Maoists."[75]

As in so many other parts of the world, Indigenous people are at the forefront of protests over coal and land dispossession in India. In the face of a state-sponsored terror campaign carried out in the name of counter-insurgency, and with legal protections like the Forest Rights Act increasingly hollowed out, Adivasi communities and other forest defenders have no option but to peacefully protest by blockading roads, sitting in front of bulldozers, and organizing rallies and marches like the ones that led to the arrests of Hidme Markam and Badri Gawde. These protests often draw on the Gandhian legacy of nonviolent direct action (known in India as *satyagraha*), which leads to campaigns such as the Koyla Satyagraha, waged against coal mining in the state of Jharkhand.[76]

A group of activist-scholars linked to the Global Atlas of Environmental Justice project in Barcelona have used the atlas's important database to analyze instances of Global Blockadia. The group's analysis finds that environmental defenders around the world "are frequently members of vulnerable groups" who "employ largely non-violent protest forms."[77] Land defense comes with a heavy toll: defenders globally face high rates of criminalization (20 percent), physical violence (18 percent), and assassinations (13 percent). This toll is particularly high in India, which has the second highest rate of murdered land and environmental defenders in Asia, after the Philippines.[78] Violence against land defenders in India also comes in many forms, including not just physical harm and murder but also cultural and ecological destruction.[79]

Despite such heavy repression, environmental defenders successfully halt environmentally destructive and socially conflictive projects in 11 percent of cases globally.[80] According to the Indian scholar-activist Brototi Roy, movements are most effective when they manage to combine strategies of resistance, including collective mobilization before projects actually break ground, protest diversification, and litigation. In these cases, Roy and her colleagues concluded, Global Blockadia is able to shut down up to 27 percent of extractive projects around the world.[81] This quite remarkable rate of success against the combined forces of capital and the state has resulted, Roy told me, from coalitions that combine frontline environmental defenders, scientists, anti-extractivist activists, and even schoolchildren from India's Fridays for Future movement.[82]

Another example of such cross-sectoral solidarity that Roy described is the group Youth Action to Stop Adani, which produced a documentary denouncing the Adani coal company's plans to expand extraction in areas of rich biodiversity. The video savvily linked the struggle against Adani to the farmers' fight against BJP efforts to ram through neoliberal reforms to India's agricultural sector. Anti-extractivist solidarity across national borders is also evident in the Stop Adani campaign, which developed after the public announcement of a $1 billion loan by the Australian government to Adani for a planned rail line in northern Australia.[83]

In an unexpected turn relating to the COVID pandemic, Adivasi resistance to land dispossession has also been intensifying in recent years, according to Jo Woodman of the human rights organization Survival International. Woodman told me that in the years before the pandemic, the government had developed a strategy of placing Adivasi youths in residential schools, where they are taught that Hindu culture is superior to their own and that *vikas*, or development through extraction of the country's mineral resources, is the way forward. These schools present a striking parallel to the abusive policies implemented at boarding schools for Indigenous children in settler-colonial states like Canada and the US.[84] As a result of the pandemic, however, many of these residential schools were shut down, and Adivasi young people returned home. There, according to Woodman's research, they witnessed the terrible cultural and environmental toll of extractivism in their communities and reengaged with Adivasi forms of creative resistance.

To strengthen these frontline forms of environmental defense, the doctrine of development can be challenged by looking at who really benefits from extraction. After all, it is not just a state like Chhattisgarh, which is represented by governing elites as "backward" and in need of electricity and jobs, but the entire country that is being turned into a sacrifice zone. In the era after independence, the coal extracted in India was not exported but rather used domestically, making it easier for officials to claim that extraction was necessary for the common good of the Indian people. It is true that overall access to electricity in both urban and rural areas improved during the period of rising coal extraction and use in recent decades. In 1990, just over 40 percent of the Indian population had access to electricity; by 2017, 85 percent of Indians lived in electrified homes.[85] This increasing proportion is, however, a result of the almost complete electrification of the country's cities (excluding slums, of course). Access remains much lower in rural areas, and, as Roy and Anke Schaffartzik document, gains in electrification are not proportional to increasing coal extraction—suggesting that "access to electricity is not functionally hinged on expanding coal extraction."[86]

An especially revealing case is that of Mundra, a small fishing village on the coast of the western state of Gujarat. In 2011, the village's Association for the Struggle for Fishworkers' Rights complained to the World Bank that the Tata Mundra power plant that had been built right next to their village was raining down toxic ash, poisoning them and the waters they rely on.[87] They claimed that the bank had refused to even acknowledge them as affected people. Tellingly, the people of the village receive no power from the enormous 4,150-megawatt coal-fired power station, the first of sixteen state-planned "ultra mega power projects," or UMPPs, that were supposed to use the most up-to-date technology to provide cheap electricity to cities in five neighboring states. The villagers took their case all the way to the US Supreme Court, which ruled that the International Finance Corporation, the World Bank's lending arm, could not be held liable for damage to the environment in and around Mundra because the damage happened in India not the US—even though the World Bank is controlled by the US.[88] To date, only two UMPPs have been commissioned, but the scheme has helped legitimate the use of imported coal (largely from Indonesia and Australia) by privately owned power plants along the country's coast.[89]

As for the idea that extraction generates jobs, India's coal sector currently directly employs a workforce of approximately 1.2 million, along with a slightly larger informal mining sector. That is significantly less than 1 percent of the Indian population, and that number is shrinking rapidly. In the immediate postcolonial era, state-owned coal companies like Coal India offered well-paying permanent jobs to families as compensation for loss of their land. But private corporations are under no such obligations, and with the sweeping privatization of coal extraction since the 1990s such private firms have become ubiquitous in India. These private companies often operate as what are known as "mine developer and operators" (MDOs)—subcontracting outfits for the national entities that control mining rights.[90] Because private companies are under no obligation to provide good jobs to the people whose land they take, coal mining jobs in India are now hard to get—and harder to keep. Mining firms tend to hire workers on casual contracts that pay a third of what state-owned companies like Coal India offer. Such casual contracts can be terminated whenever the employer wishes. Even Coal India recently announced that it would outsource its mining operations to private MDOs.[91] To make matters worse, contrary to its public image of masses of dust-encrusted workers hewing into the bowels of the earth, mining has become one of the world's most highly automated industries, employing a minimal workforce supervising robot trucks, drills, shovels, and trains.[92] Opencast mines, which slice off the earth's surface (the industry revealingly calls it "overburden"), employ only a fraction of the workers once employed in pit mines.

Union leaders such as Sudarshan Mohanty, who represents mine workers in Odisha, argue that India needs a strategy to transition from coal to cleaner energy.[93] It is indeed tragic that people displaced by mines in India's coal belt are now dependent on mining. In 2018, the Indian government announced a coal phaseout, a position that it reiterated at the United Nations climate conference in 2021. What this might mean given the 2020 plan to "unleash coal" remains unclear, as do any specifics about how those dependent on the coal industry will be transitioned to other jobs. In a 2021 report on how to craft a just transition for Jharkhand, one of India's most devastated states, where underground coal fires have been burning for a century, energy researcher Sandeep Pai urges the government to grow non-fossil-based sectors such as agriculture, tourism, and renewable energy.[94]

While well-intended, these suggestions seem rather naive since they skate over the environmental devastation caused by mining, which is likely to seriously compromise industries such as agriculture and tourism. Moreover, the renewable energy industry in India, like coal extraction, has been criticized by activists for exacerbating land dispossession and denying rural communities access to the energy generated on their land.[95] It seems clear that the conversation on a just transition, just beginning in India, has a long way to go, and has much to learn from countries where it has been a site of intense debate and political contestation for some time. For Ulka Kelkar, director of climate at the World Resources Institute's India chapter, India needs to look to South Africa for inspiration about what a just transition could look like.[96] Accordingly, it is to the prospect of just transition in South Africa, one of the world's most unequal societies, that I now turn.

## "If Need Be I Will Die for My People": The Toll of the Minerals-Energy Complex in South Africa

On the night of October 23, 2020, four men broke into the home of Mama Fikile Ntshangase in the town of Ophondweni, KwaZulu-Natal Province, and shot her dead. Mama Fikile, a sixty-five-year-old activist in the fight against expansion of an opencast mine operated by Tendele Coal near Hluhluwe-iMfolozi Park, the oldest nature reserve in Africa, had received multiple death threats in the past. When asked about these threats shortly before her killing, Mama Fikile said, "I cannot sell out my people and if need be I will die for my people."[97]

The assassination of Mama Fikile is part of a pattern of mounting violence toward environmental activists in South Africa and beyond. Her killing was preceded by the 2016 murder of Sikhosiphi Rhadebe, a campaigner for the rights of residents of the Xolobeni community in the Eastern Cape and an opponent of a proposed opencast titanium mine operated by Mineral Commodities Ltd., an Australian mining company, which was slated to be developed in the Wild Coast region of the country. These assassinations of two courageous individuals are part of a broader campaign to silence opposition to mining. As the rights organization Global Witness has documented, the mounting numbers of assassinations of environmental defenders each year suggests that violence against those protecting their land and the planet increases in direct proportion to the intensification of global environmental crises.[98] These killings are part of an undeclared but nonetheless vicious war on environmental defenders, one in which a panoply of brutal tactics are deployed to quash communities' efforts to protect their land and their lives.

Mama Fikile's killing is a reminder that women often play a leading role in anti-extractivist activism. The gendered division of labor in many communities assigns the essential tasks of social reproduction, such as providing food, to women. These tasks depend on a stable environmental foundation. Since extractivism expropriates and pollutes local land and water, women tend to be on the front lines of resistance. The harassment meted out to activists therefore often has an element of misogyny and gender violence.[99] Many times, women are reluctant to report such threats, fearing that they will be shamed for having been targeted with sexual violence either by the perpetrators themselves or by patriarchal forces within their own communities.

Threats of violence against anti-extractivist activists create an environment of fear that has a chilling effect on entire communities. As UN special rapporteur on the situation of human rights defenders Mary Lawlor argues, threats—both direct and indirect—sometimes target individuals and sometimes menace entire communities, but they always constitute a "direct attack on civil society space."[100] As Lawlor notes, statistics on the number of killings are often not compiled nationally, and information about death threats is even harder to come by.[101] Nor are assassinations the only mode of violence employed by extractivist corporations and their allies in government. A 2019 report published jointly by Human Rights Watch and three South African environmental organizations documented how activists in mining-affected communities across South Africa experienced not just death threats but also physical violence and property damage in retaliation for their activism.[102]

People also face intimidation when they try to mobilize collectively in public, often from the police, who violently disperse protests and arrest protesters arbitrarily.[103] Finally, companies often try to silence activists using the court system, filing "strategic lawsuits against public participation," or SLAPPs, which aim to censor, intimidate, and muzzle critics with expensive legal proceedings. In 2021, the South African High Court handed environmental defenders and free speech advocates a victory by holding that a series of defamation lawsuits brought by an Australian mining company were an abuse of the legal process.[104]

In addition to these frontal assaults on individuals, communities, and on civil society more broadly, mining companies also engage in forms of what cultural critic Rob Nixon has called "slow violence"—the gradual poisoning of the air, earth, and people's bodies whose results often take decades to achieve full impact.[105] In KwaZulu-Natal, scarcely a year after Mama Fikile's assassination, the breaching of a pollution control dam at a coal mining facility owned by a multinational mining subsidiary led to a flood of contaminated water flowing into local rivers in the area around the Hluhluwe-iMfolozi Park.[106] This contamination crisis was only one instance of coal companies' environmental destruction in the area. In 2017, the South Africa–based anti-extractive feminist organization WoMin released a report based on participatory action research with women in the area around the Tendele coal mine. It documented the twin crises caused by the mining company's water grabs and a climate change–exacerbated drought.[107] In an area in which women-headed households rely on crop and livestock farming to supplement remittances from male migrants to industry and mines in other parts of the country, the arid countryside was strewn with animal carcasses and skeletons as the drought wiped out the livestock in which rural wealth was vested. Women from nearby communities accused the Tendele mine of water grabs as well as air, water, and soil pollution that destroyed local farming, saying that the mine was pumping water from the local uMfolozi river and fencing off communal water sources.

The region of northern KwaZulu-Natal where these conflicts took place is an area of great ecological significance, but it is also a region of intense extractivism, located just an hour's drive away from the Richards Bay coal terminal, the largest coal export terminal in Africa. The Hluhluwe-iMfolozi Park, a hunting grounds for King Shaka and previous Zulu kings for over 250 years, is almost entirely encircled by mines. Once-thriving farming communities in the area have lost their land as the mines have polluted the surrounding air, water, and land, and are dependent on drinking water trucked in by the local municipal authority. Residents now live on the edge of annihilation. As a woman named Zandile from the town of Somkhele put it, "How are we supposed to survive without water? We are dying, our animals and crops are dying. We cannot continue like this much longer."[108]

Rural KwaZulu-Natal is not the only sacrifice zone in South Africa. The country is quite literally grounded on extractivism, and consequently is pervaded by the spaces of social abandonment and ecocide associated with extractive operations. Colonial conquest by Dutch and British settlers led to large-scale land dispossession and to the growth of monocultural plantation agriculture. With the discovery of diamonds, gold, and coal in the mid-nineteenth century, capital and the state in South Africa intensified land enclosure to create a migrant labor system in which Africans migrated to work in the mines from the "reserves" (later Bantustans) to which they had been consigned by settler colonialism. With its barrage of racist legislation, the apartheid system intensified the lopsided character of the country's economy, focusing development on a mining-based economy dependent on the abundant Black labor produced by forced removals and what was in effect a forced labor system.

South Africa's economy was dependent on "cheap" fossil fuels.[109] The country's Minerals-Energy Complex (MEC), a label established by economists Ben Fine and Zavareh Rustomjee, was able to develop thanks to tight integration of the apartheid state and mining conglomerates like the massive Anglo American Corporation.[110] Founded in 1923, the state-owned utility Eskom generated electricity by burning coal dug up from Anglo American's mines. Most of this state-generated power then went to fuel Anglo American's gold and diamond mining operations, with residual capacity being used to furnish electricity for white consumers. The vast majority of the country's Black population was completely cut off from the grid. The state-owned Iscor iron and steel corporation developed manufacturing sectors of use to mining interests, again using state-generated power, but industrial production not directly connected to the MEC remained in a state of underdevelopment throughout the apartheid era. Given the scale of funding required by the country's mining operations, finance capital developed as an essential ally of and subordinate to the country's big extractive conglomerates.

Extractivism has ravaged South Africa's landscapes and people, and not just in the northern KwaZulu-Natal region where Mama Fikile lived. In response to the ruination generated by mining, a group of grassroots environmental organizations formed the Push Back Coal Coalition. Their 2018 report, entitled *Coal Kills*, documents the impact of the coal industry on the Mpumalanga Highveld, the heart of South Africa's coalfields.[111] A fertile region in the northeast of the country, the Highveld is home to 54 percent of the country's viable agricultural land and is the source of much of the nation's fresh water supply. Nevertheless, the coal industry has polluted the soil, water, and air in the region, depleting arable land and ecosystems and generating widespread displacement. Acid is draining out of abandoned as well as active mines, as pyrites in the rock turn into sulfuric acid after encountering oxygen, a chemical process which in turn mobilizes heavy metal toxins, poisoning water supplies and turning the Highveld into a wasteland.[112]

One of the most fecund places in the country has been turned into a toxic hellscape, with fires breaking out spontaneously in mines, on coal stockpiles, and on discard dumps, filling the air with volatile organic compounds. In addition to the underground ruination, the Highveld is also home to twelve of Eskom's coal-fired power plants, which dump millions of tons of toxic gases into the air every year. The region is also the site of the world's single largest source of carbon dioxide, the Sasol corporation's Secunda plant, a facility built to turn coal into oil to help the apartheid government evade an international oil embargo against the racist regime.[113]

Despite all the power being generated for industry, energy poverty affects roughly half of all households in South Africa.[114] Around two million people are unable to afford electricity, even when it's available, and instead burn coal or paraffin for domestic energy, thereby exposing themselves to toxic fumes indoors, as well as to a high risk of fires that can rip through informal shack dwellings like wildfire. If South Africa's poor thus get doubly exposed to airborne toxins, mineworkers face a triple threat. But coal mining and power generation provide 5 percent of formal, well-paying jobs in Mpumalanga: those workers—and the people of South Africa more broadly—need a viable alternative, or they will resist efforts to dismantle the MEC.[115]

Although South Africa's economy changed after the end of apartheid in 1994, it has not fundamentally transformed. The result has been a deep structural crisis that is both socioeconomic and ecological. Once in power, the ANC largely abandoned the socialist promises of the foundational Freedom Charter and any interest in wealth redistribution, instead adopting a neoliberal economic framework that included voluntary trade and financial liberalization. This orientation undermined the already-skewed production structures of the MEC, generating widespread deindustrialization and a relative shift of the economy toward financialization.[116] For example, the contribution of manufacturing to South Africa's GDP declined 44 percent from 1990 to 2018.[117] Once a cornerstone of industry in the country but now privatized and exposed to devastating international competition, steel production collapsed, taking thousands of jobs with it.[118] In tandem with this, economic liberalization after the arrival of democracy facilitated large-scale capital flight. Financialization meant that domestic capital shifted away from the kind of long-term investment necessary to diversify the country's industrial base, and toward finance and consumption and sectors with strong links to such activities.[119]

These developments exacerbated injustice in South Africa, which is now officially the most unequal society in the world, with the poorest 20 percent of South Africans receiving only 1.6 percent of total income while the richest 20 percent earn 70 percent.[120] White domination and ownership of the economy remains largely intact, despite the creation of a small class of Black economic and political elites through post-apartheid "Black Economic Empowerment" policies. Meanwhile, one-third of South Africa's workforce is unemployed, those with work are sinking deeper into debt, and the average life expectancy of Black South Africans is below fifty years.[121]

South Africa's post-apartheid trajectory has intensified its dependence on mining and related extractive industries, making it more of a carbon nation than before. But although it contributes 8–10 percent of GDP, the country's extractive sector has all the signs of an expiring industry. In 1990, there were 780,000 South African mineworkers; today there are less than half a million. There has been a sharp fall in the number of active, publicly listed mining companies on the Johannesburg

Stock Exchange and an utter collapse of exploration spending.[122] Existing mines are being exploited to the point where they're no longer economically profitable, and new ones are not being built.[123] Although there has been an uptick in industry profits as a result of the post-pandemic commodities boom, the failure to open new mines in the country is a recipe for terminal decline.[124]

Analysts give a number of reasons for the industry's dire straits despite South Africa's unparalleled mineral wealth. These include patently reactionary explanations such as "labor unrest"—i.e., militant demands from the country's mineworker unions.[125] Mine owners have a history of brutally crushing South African workers: even as apartheid was coming to an end, for example, a strike led by the recently formed National Union of Mineworkers was quashed, setting the terms for neoliberal restructuring of the country's labor force, which facilitated the shrinking of permanent workers and the growth of precarious labor.[126] The Marikana Massacre of 2012—in which thirty-four striking miners were shot dead by the police—shocked the country and demonstrated the toll of an exploitative migrant labor system that remains integral to the industry's outsize profits.[127] Job-killing automation has been one of the industry's primary responses to worker militancy, a global trend that has transformed mining into one of the most fully robotic industries in the world.[128]

But analysts also cite logistical problems such as the crisis of Eskom, which by the early 2000s went from having excess generational capacity to a shortfall of power, leading to increasingly expensive electricity, to the unpredictable power outages known as "load shedding," and to cuts in power supplies to large customers like mines and mineral processing plants.[129] Whatever the cause, the Minerals-Energy Complex, the mainstay of the South African model of extraction and capital accumulation, is breaking down. South Africa must now build an alternative economy that is not founded on the ruinous model of extractivism.

## The Fight for a Just Transition in South Africa

The service delivery protests that began in 2001 were an expression of popular rage at the enduring injustices of the South African economy, which very much remains grounded in the Minerals-Energy Complex that defined the apartheid era. In 2011, a coalition of unions and civil society organizations went beyond such protests by mobilizing to provide an alternative economic model during the run-up to COP17, the UN Climate Summit held in the South African city of Durban. The intention was to overcome the reactive and relatively fragmented politics that characterized "service delivery protests" and articulate a vision for a just transition to environmental and social sustainability in South Africa.[130] Taking the lead in this effort was the National Union of Metalworkers of South Africa (NUMSA), which at the time was a member of the federation of trade unions known as COSATU. Before COP17, NUMSA's energy research and development group met with climate scientists and held workshops on decarbonization with workers to come up with radical solutions to South Africa's compounded crisis. These discussions led to the production of the *Policy Framework on Climate Change*, the first publication organized around the theme of a campaign for a "million climate jobs" in South Africa.[131] The idea of a just transition was central to the calls to action that emerged from the union movement and civil society organizations.

The term *just transition* emerged from the US labor movement in the face of the slated closure of a large chemical facility in New Jersey in the mid-1980s, which led the Oil, Chemical and Atomic Workers Union (OCAW) to fight for a "Superfund for Workers."[132] Rather than simply struggling for income protection for the plant's workers, in other words, the union pushed for a program of government-funded retraining for the workers who would be displaced by the plant's closure. OCAW President Tony Mazzocchi coined the term *just transition* in this context to argue that workers should be compensated for policies of environmental protection that threatened their jobs.

This idea was taken up by South African unions such as NUMSA and COSATU in their debates about how to cope with the climate emergency. COSATU's *Policy Framework on Climate Change* centered the concept of just transition, calling for the participation and leadership of organized labor in the country's climate change policy discussion and requesting government support for renewable energy and climate jobs.[133] This stance helped generate a national dialogue about the climate emergency that is among the most progressive in the world. South Africa was, for instance, the first country to include mention of a just transition at the UN Climate Process, at the 2015 COP21 conference in Paris.[134]

But the capitalist class has tried to co-opt the idea of a just transition, emptying it of its insistence on a class struggle to redistribute power and resources. For this reason, General Secretary of the South African Federation of Trade Unions Zwelinzima Vavi argues, the climate justice and trade union movements must insist on "a profound socioeconomic transformation" as the "only way to achieve 'a zero-carbon world.'"[135] For Vavi and radical South African unions such as NUMSA, this means that fossil capitalism can only be overcome by a workers' struggle framed around a "socially owned renewable energy sector."[136] In its 2012 declaration, NUMSA argued in forceful terms that the urgency of the climate crisis should not lead activists in South Africa to forget questions of public power and democratic ownership of renewables. As NUMSA put it:

> As energy prices go through the rooftop, as energy poverty/inequality continue to persist, as communities and a number of countries rapidly lose their energy sovereignty and the right to determine their energy choices, as the right to energy remains a dream for millions of the globe's citizens, as state-owned energy enterprises continue to act like private energy companies; the environmental, political, and economic case for genuine public ownership and democratic control of energy is becoming more strong.

Compare NUMSA's position on public renewables to the most progressive policy document published on a national level in the US. The Bernie Sanders–backed Clean Energy Worker Just Transition Act calls only for federal investments in renewables, a far less radical proposal than the fully nationalized, democratized, and equitable power sector demanded by NUMSA.[137]

The campaign organized by NUMSA, affiliated unions, and civil society groups like the Alternative Information and Development Centre focused on three key sectors for just transition and job creation: renewable energy, the built environment, and transportation. In a booklet entitled *One Million Climate Jobs*, the campaign set out its vision for "moving

South Africa forward on a low-carbon, wage-led, and sustainable path."[138] It is worth noting that this booklet was published in 2016, well in advance of the most recent wave of Green New Deal manifestos published in the US and Britain.[139] The South African plan and the union mobilization that helped produce it offer a blueprint for what a just transition could look like in other countries, including other Global South nations as well as core imperial ones like Britain and the US. The AIDC writers calculated that in order to slash its carbon emissions in line with the global goal of limiting warming to 1.5° C, South Africa would have to massively speed up deployment of renewable energy.

According to prominent South African energy scholar Mark Swilling, the country is a "well-known early mover" in adopting renewable energy, having established a system in 2011 through which energy companies would bid for the right to install renewables, the theory being that competition between companies would drive the cost of renewable energy down.[140] As a result of this process, known as the Renewable Energy Independent Power Producer Procurement Programme (REI4P), 112 projects have been set up over the last decade, generating a total of 6.3 gigawatts from modern renewables like solar and wind, with another 6 gigawatts planned. This is far more renewable power than many countries have commissioned. Yet, as the AIDC points out, this total is nowhere near enough, because if all this power were actually up and running and connected to the grid by Eskom, it would still leave 88 percent of South Africa's electricity coming from coal.[141] Another critical issue flagged by the AIDC is that the REI4P program explicitly states that "45 percent of value will be provided by work done in the country," which means that almost half the jobs will be located outside South Africa—a huge problem for a nation facing cripplingly high unemployment.[142]

In place of this anemic market-led effort, the AIDC argues that South Africa needs a massive public energy transition program. The scale of this program must be huge since the country needs to triple energy production in order to electrify everything—not just energy generation but also transportation and the heating and cooling of buildings.[143] The *One Million Climate Jobs* proposal is to install 6 gigawatts of wind and 9 gigawatts of solar power *every year, for twenty years* in order to eliminate coal power from South Africa's grid. And the grid itself would have to be extended and updated, turned into a "smart grid" that uses computers to balance the variable amounts of energy generated by renewables.

AIDC writers calculate that this would generate 250,000 jobs, all of which would be permanent and half of which would be in manufacturing.[144] Since these jobs would all be public, the government could ensure that they would all be well-paying, secure, and that women would have equal access to them. This would overcome one of the major hurdles to energy transition: the resistance among unions around the world based on the precarious, ill-paid character of work in the predominantly private renewable energy sector. The promise of good work is a key element of just transition, but it has been sorely lacking in most policymaking and discourse. Instead of thinking about the issue simply in terms of fossil fuels versus renewables, South Africa's labor unions and allied groups like the AIDC are thinking about people's power as the foundation of the transition.

In addition to the quarter million jobs to be created through direct production of the new power infrastructure, AIDC writers calculate that nearly ninety thousand jobs would be generated in maintenance and repair of this infrastructure. Many more jobs would also be produced in ancillary industries such as construction and engineering. But, as we have seen, the energy sector is not the only one that must be completely remade as part of the just transition. Transportation is an increasingly significant source of carbon emissions, responsible for almost one-third of final energy demand globally. Transport emissions from developing countries are expected to nearly triple by 2050.[145] Owing to South Africa's extreme economic and social inequality, the majority of people in the country don't currently own private automobiles. But, if and when South Africa becomes less stratified, there needs to be a plan in place to ensure that people don't acquire cars and thereby lock the country into unsustainably high levels of emissions from the transit sector.

To this end, the *One Million Climate Jobs* proposal calls for the creation of enough new buses, trains, and minibuses for eight million people—all of which would need to run on clean electricity rather than dirty diesel fuels (95 percent of buses still use diesel, even in Europe).[146] But it's not simply a case of ensuring equitable access to clean public transportation.

Just transition efforts in the transportation sector must also ensure that public transit is not just electric but also efficient and affordable, thereby increasing mobility justice.[147] This means that special lanes should be set aside for a Bus Rapid Transit system, which can whisk travelers past drivers sitting gridlocked in traffic. The AIDC calculates that this new public transit fleet would add 370,000 jobs each year, for twenty years. In addition, to ensure that dirty vehicles are gradually taken off the road, the AIDC argues that South Africa needs a law declaring that all new car and truck sales must be electric within five years. With fast trains connecting the nation's cities and a convenient bus service, polluting domestic air traffic could be largely shut down.

The last major area tackled by the *One Million Climate Jobs* proposal is carbon emissions deriving from housing. Here we see significant overlap with the strategies of housing activists profiled in the previous chapter. What's novel in this case is the link between housing issues and energy poverty and injustice. As AIDC writers note, South Africa's building-based emissions are currently quite low, constituting only 5 percent of emissions. This is because of the country's relatively mild climate, but also because so many of its residents cannot afford electricity. We've already seen that this poverty leads to high mortality levels based on the burning of paraffin and coal indoors. The AIDC report notes that there is huge pent-up demand for decent, affordable houses, and suggests that municipalities throughout the country should be tasked with mapping both communities most at risk from climate change and communities lacking basic services like electricity, water, and sewage.[148] According to AIDC estimates, two hundred thousand jobs could be generated by building new homes as well as retrofitting existing ones with insulation to make them as energy efficient as possible and resilient to the increasing extremes generated by the climate emergency. The work would begin with low-income homes and public-sector buildings and then expand to include the construction of food gardens, public parks, and other public amenities across the country. Individual homes would have solar water heaters and embedded photovoltaic arrays installed on their roofs, rainwater harvesting tanks, and excellent insulation.

Critics often ask where all the money for these millions of climate jobs would come from. According to the AIDC's estimates, the total cost of a national just transition plan would be 8 percent of South Africa's current economy. This is a substantial figure, but thinking about the cost of *not* embarking on such an ambitious response to the climate emergency helps put the total sum in perspective. As the AIDC puts it, "Affordability is a matter of choice. From an economic perspective, it is never 'expensive' to a nation to organize, educate, and train previously unemployed people in work that creates public assets and productive wealth, like wind and solar power stations, or public transport facilities."[149] As they remind readers, countries seldom have problems finding money to bail out bankers or buy weapons when a financial or military emergency is declared. The climate emergency is far deeper than the crises provoked by capitalism's periodic boom-and-bust cycles, a fact particularly apparent in South Africa, one of the world's climate change hotspots.

The *One Million Climate Jobs* proposal argues that action must be taken, and that the resulting just transition would not only create valuable low-carbon infrastructure but also economic growth that the government could marshal. As a result, the AIDC calculates that the government could recoup and recirculate two-thirds of the money invested in the project. The other upfront costs could be covered, the proposal argues, by taxing corporations and high-income South Africans, by closing tax haven loopholes, and by using money stored in government unemployment insurance and pension funds.[150] The result would be a massive transformation in South Africa's built environment and society. Beyond simply the green infrastructure and the one million direct and seven hundred thousand indirect jobs that would be generated, the AIDC contends that its proposal would give South Africans a sense of collective agency:

> When South Africans took on the apartheid system and won, the whole world saw. If South Africans win a million climate jobs, workers and ordinary people all over the world will notice. They will see us on their televisions, and come to visit and learn. They will understand that there is a way of dealing with climate change, and there is a way of dealing with austerity.[151]

The One Million Climate Jobs campaign thus invokes the world-historical significance of the South African people's relatively recent victory against apartheid, one of the last and most vicious forms of formal racial segregation in history. The South African people can once again fire the world's imagination and become a beacon to other nations, the campaign argues, by winning the fight for a just transition. A victory of the South African people against the terminal trajectory of fossil capitalism would be an inspiring example of environmentalism from below. As the AIDC notes, "There is much at stake—the future of South Africans, and the future of humanity."[152]

If a just transition requires the thoroughgoing material transformation of a country such as South Africa, even more crucially, the AIDC writers note, it will require a deep political upheaval. This prediction proved all too prescient. Although, as the AIDC's Sandra van Niekerk told me, the One Million Climate Jobs campaign in 2014 organized a successful drive to collect one hundred thousand signatures from people in support, internal discord between South Africa's unions soon shifted focus away from the campaign.[153] This discord was a product of conflict between those who believed that the South African labor movement should maintain its long-standing alliance with the ruling ANC party, and those who believed that workers needed a more independent, militant federation. The latter group included NUMSA leadership. This conflict led to the expulsion of NUMSA, the core union behind the climate jobs campaign, from the union federation COSATU in 2015. But even if these tensions momentarily shifted attention, the climate crisis was not going away: in 2017, Cape Town's nearly four million people came terrifyingly close to running out of water, a situation that was referred to as "Day Zero."[154] The union-affiliated research institute NALEDI reacted by spearheading workshops around climate change and the issues confronted by trade unionists.[155]

In tandem with these organizing efforts, NUMSA continued to affirm its vision of a socially owned renewable energy sector. In this way, it offered an alternative to the privatized Renewable Energy Procurement Programme instituted by the ANC government.[156] As NUMSA deputy general secretary Karl Cloete argued in an op-ed, that program is "not meant for those who do not presently have access to electricity but is being developed for big corporations which get their supply at a discount."[157]

It was clear by this point that the struggle for socially owned renewable energy in South Africa hinged on resisting the ANC's plans to privatize Eskom. NUMSA and other unions joined this fight through three months of rolling strikes in 2018. The following year, radical unions again took to the streets to demand a just transition instead of the neoliberal one that the ANC foisted on the nation. And in 2021, NUMSA marched to Eskom's headquarters in Johannesburg to oppose the government's plans to privatize the utility.[158]

These demonstrations made it clear that for union activists, the road to public power—in multiple senses of the term—has to run through Eskom. But Eskom is in a terrible state, deep in debt and struggling to maintain services. Once capable of producing some of the cheapest and most plentiful power in the world, Eskom now struggles with power shortages that lead to rotating blackouts, undermining the country's economy, cutting into the utility's revenues, and, as Trevor Ngwane and his colleagues argue, disproportionately impacting the poor.[159] Rather than cutting off individuals, Eskom shuts power off to whole zones of cities and even to entire rural towns through what is referred to as "load reductions."[160]

The popular press continually links Eskom's woes to the ruling party's corruption, pointing in particular to the failure to complete two huge new coal-fired power plants—Kusile and Medupi—as examples of "state capture" by corrupt elites linked to the ANC.[161] But while corrupt kickback schemes are undeniably a factor in the crisis of the state utility, Eskom's woes are more importantly linked to structural factors that have pushed many of the world's utilities into debt and crisis. In 2020, the AIDC joined with two organizations, New York–based Trade Unions for Energy Democracy and the Transnational Institute in Amsterdam, to research and publish the report *Eskom Transformed*. According to their analysis, the utility's problems "are in fact a direct consequence of the 'electricity for profit' paradigm promoted by the World Bank and other neoliberal institutions."[162] Eskom's purported death spiral is therefore characteristic of the parlous situation of many public utilities around the world. Understanding the crisis of the public utility is thus a critical political struggle. Figuring out how to transform a public utility like Eskom into a vehicle for socially owned renewable energy and public power is a key terrain of conflict, one with implications that go far beyond the fate of a single nation.[163]

Eskom's crisis is particularly tragic given the high hopes radicals had for it as a central vehicle for realizing the liberation struggle's promises of equality and inclusion. Shortly after the end of apartheid, the Eskom Amendment Act in fact expanded government control of the utility and articulated an aspiration for universal access to electricity in the country.[164] In 2003, the Free Basic Electricity program was introduced, prompted in part by the militant protests of the Soweto Electricity Crisis Committee and similar groups. Today, over 88 percent of South Africans have access to electricity, a startling success given that only 40 percent of the country had such access at the end of the apartheid era in 1994.

In tandem with these efforts to expand access to power, however, the ANC embraced policies pushed by the World Bank and other international financial institutions to "unbundle" Eskom, or to break up the public utility into separate generation, transmission, and distribution units. The argument was that despite the increased transaction costs associated with such fragmentation of the utility, unbundling would allow greater market participation of the new units, generating competition that would make the separate units more economically competitive.[165] As South African energy analyst Brian Kamanzi points out, the idea of unbundling emerged in the 1970s, when neoliberal ideologues argued for the dismantling of large state companies, leading to electricity-sector reforms in Chile under the Augusto Pinochet dictatorship and in Britain under Margaret Thatcher.[166] Unbundling was supposed to unleash competition that would lower costs for electricity consumers and produce more nimble, efficient energy provision. Instead, it has led to profit-seeking that has stripped essential infrastructure to the bone. The result has been increasing grid instability, as the deadly power outages in Texas during the winter of 2021 demonstrated.[167]

The expectation that Eskom would be chopped up led many in South Africa to regard the utility as a kind of zombie, an entity that was effectively dead though it still lumbered around the country. This impression was strengthened when the ANC government instructed Eskom to cease building new capacity and to sell off generation assets to ensure 30 percent private ownership by 2004.[168] Pushback from unions has led the ANC to repeatedly delay its unbundling plans, although the announcement of an $8.5 billion debt-relief package for Eskom at COP26 in 2021 included an explicit directive to continue "policy reform in the energy sector," including "unbundling and improved revenue collection."[169]

The primary avenue for the privatization of South Africa's energy system has lamentably been the renewable energy sector—a fact that helps explain NUMSA's criticism of the COSATU-ANC alliance and its plans for national energy transition. In 2009, the South African government introduced a system of feed-in tariffs modeled on European policies, a scheme that provided long-term guaranteed subsidies for anyone that fed renewable energy into the electric grid. In Europe, feed-in tariffs incentivized ordinary citizens and communities to invest in small-scale solar, biomass, and wind generation for their homes and local regions, leading to a remarkable expansion in renewable energy production.[170] But despite the manifest benefits of citizen participation, some in government saw the feed-in tariffs as too generous, and as generating political backlash from consumers who saw their taxes go up to pay for the energy transition.

South Africa was an early adopter of an alternative system based on auctions in 2011. Promoted as a scheme that can ensure a lower cost for renewables through a competitive bid process, the Renewable Energy Independent Power Producer Procurement Programme, or REI4P, has been hailed as a success story that other countries of the Global South should copy. Yet the total amount of wind and solar power installed in South Africa from 2013 to 2019 was only 3.9 gigawatts, or just 3–4 percent of the country's electricity.[171] What's more, the entire system relies on public subsidies to make the renewables sector appear competitive with fossil fuels and hence attractive to investors, according to the authors of the *Eskom Transformed* report.[172] The public thus essentially bankrolled guaranteed profits for private investors.

This system is quite clearly inequitable since everyone has paid for renewables but only some can afford to install them. As it is deployed in South Africa, the upshot is a worsening of climate apartheid, where the wealthy buy amenities that soften the impact of the climate crisis while the poor struggle to survive conditions generated by the consumption of the rich. In addition, since it guarantees favorable returns for investors in renewables, the auction system also tends to generate a profitability crisis for utilities. This is because utilities are mandated to buy power from private renewables companies, even while they continue burning increasingly expensive fossil fuels to run the grid.

But why do utilities keep using fossil fuels? Part of this has to do with so-called stranded assets, the fossil-based infrastructure that utilities have invested in over the years and that in a sense holds them hostage. As renewable energy becomes more affordable and widespread, fossil fuel–based infrastructure loses more and more value despite the huge investment made in its construction. But utilities also face technical challenges related to energy transition. Because the wind doesn't always blow and the sun doesn't always shine, utilities charged with balancing the grid to keep it from crashing must use fossil fuels to generate what is known as "base load" power, which can be deployed quickly when the variable renewable power in the grid drops. Huge amounts of public money are spent keeping fossil-based plants sitting idle, waiting to ramp up base load power.[173] Coping with increasing quantities of variable power is also increasingly costly and technically challenging, but since public funds are going to private renewable companies, the utilities are caught in a kind of death spiral where they are charged with maintaining expensive infrastructure while facing ever-diminishing investment.

But perhaps the most important and even fatal flaw in the auctions-based arrangement that inspired South Africa's REI4P is the assumption that the cheapness of renewables will kill fossil fuels. The wager here is that cheap renewable power will inevitably lead to a tipping point where investment shifts away from fossil fuels and to renewables, leading to a quick, market-driven energy transition. This narrative is hegemonic in official policy forums around the world.[174] But, as we have seen, fossil fuels are not dying but rather are expanding *in tandem with* the introduction of renewable energy.

In addition, as the authors of *Eskom Transformed* argue, as bid prices for contracts to generate renewable power have fallen, investors see their outsize profit margins in the sector evaporating. This inevitably leads to lower levels of investment, as capital seeks out bigger returns elsewhere.[175] Deployment of renewable energy has in fact basically plateaued since the middle of the last decade.[176] Given the gravity of the climate crisis and the need for a rapid transition, staying in place is essentially committing to annihilation. As the Climate Policy Initiative put it in an assessment of global climate finance, "There is a need for a tectonic shift beyond 'climate finance as usual.' Annual investment must increase many times over, and rapidly, to achieve globally agreed climate goals and initiate a truly systemic transition."[177]

In line with arguments that South African unions like NUMSA have made for socially owned renewables, the activist-scholars behind *Eskom Transformed* demand a reversal in the current trajectory toward the dismantling of Eskom.[178] Eskom must be de-marketized and rebuilt, they argue. The Independent Power Producer system that is throttling the public utility must be dismantled within South Africa, and the socially owned renewables system long demanded by activist unions must be built as the protagonist of a just and democratic energy transition in the country. In tandem with these struggles, fights for a public good approach to energy need to be waged on a global level to counter the auction-based, free-market system backed by institutions like the World Bank.

The impact of the report and of the campaign for socially owned renewables is still unclear. South African news media are filled with denunciations of Eskom, attacks which almost uniformly blame the utility for power cuts that, as we have seen, are to a significant extent a result of government acquiescence to privatization. Nonetheless, according to report coauthor Sean Sweeney, COSATU has picked up on the arguments made in the report and used them to buttress its blueprint for a just transition.[179] In preparation for COSATU's annual congress in 2022, a resolution was being circulated calling for a united front against the privatization of Eskom and for a public utility pathway. Despite his condemnation of Eskom's energy racism, Trevor Ngwane told me that he too is in favor of rebuilding the utility and fighting for a just transition.[180] While these activist campaigns for energy democracy face strong headwinds, the for-profit energy sector is clearly failing the nation. As blackouts roil the country, the need for a viable public alternative couldn't be clearer.

It is not just South Africa that is experiencing this shortfall of investment in renewables. A 2022 report by the International Renewable

Energy Agency (IRENA) describes a "fundamental disconnect" between existing energy systems and the technical and economic aspects of renewables, and as evidence cites the decline in electricity prices in places with high penetration of renewables.[181] Rather remarkably for a global policy institution, IRENA advocates an end to the capitalist economy's pursuit of ceaseless growth, and an emphasis on global equality to redress the historical and enduring impacts of colonialism and imperialism:

> A steady-state economy that properly addresses distributional aspects seems an appropriate goal for human activity on a planet that has finite resources and impact-bearing capacity. . . . Beyond solidarity, in the current climate crisis addressing the distributional dimension has become a must. Transitional dynamics in an unfair world would lead to much of the world's population getting access to very cheap fossil fuels and related technologies, because of the reduced demand for these in the global North. This could easily reduce any decarbonization advancements in the global North as the rest of the world seeks to replicate its fossil fuel–based economic growth of the past decades.[182]

IRENA's logic in this passage is notable: the agency admits that tackling inequality is a moral imperative, but then also argues that the survival of humanity essentially relies on rejecting the competitive pursuit of capital accumulation and ensuring a viable pathway to renewables for all. Energy transition approaches grounded in free-market dogmas must be abandoned, for if what IRENA describes as "the distributional dimension" is not dealt with, Global South nations will see no reason not to access and burn the fossil fuels abandoned by the North, thereby consigning us all to climate catastrophe in the long run.

These questions are more pressing than ever in a country such as South Africa. It is worth remembering that although the activists of the Soweto Electricity Crisis Committee (SECC) were successful in pushing the government to follow through on its promises of Free Basic Electricity, this program was not universal and it came with serious strings attached: households were evaluated economically to determine whether their incomes were low enough to qualify for free power, and those who did were required to register and agree to installation of a pre-payment meter in exchange for a relatively meager allotment of 50 kilowatt-hours per month of free power. Many township residents perceived the pre-payment meters as onerous given the difficulty of obtaining cards to pay for electricity. The meters were also seen as a means to throttle the government's allocation of free electricity.[183] They thus became yet another instance of the state's broken promises, tangibly present in people's daily lives.

As a result of these draconian policies, the forms of electricity commoning pioneered by the SECC have become widespread in recent years, with an estimated 85 percent of Sowetans informally reconnecting themselves to power during the COVID pandemic.[184] These popular strategies are ensuring that the anti-apartheid chant "Amandla! Awethu!" (Power to the People!) is more than simply a slogan. But unless Eskom is saved and transformed into a vehicle of just transition, this power will not be sustainable.

## Toward Global Energy Democracy from Below

India and South Africa are two of the world's most vibrant democracies, and also happen to be situated in areas of the globe most vulnerable to climate change. The fact that these two countries are among the world's most fossil fuel–dependent states might seem counterintuitive given their political and environmental circumstances. We are, however, familiar with the legacy of the Minerals-Energy Complex and histories of racism and disdain for the people whose land is taken and whose health is destroyed by coal extraction. We have seen how these histories have facilitated "development" of fossil fuels. Ideas of national energy sovereignty and the right to autonomous development have in fact prompted both countries to intensify coal extraction in recent years, notwithstanding scientific evidence about the deadly outcome of such policies. Where coal production was concentrated in the Global North several decades ago, in both core countries like the US and Germany as well as in Soviet bloc countries like Poland and Russia, today the largest number of jobs in the coal sector are in China, India, and Indonesia.[185]

Given how deadly coal is in terms of carbon emissions, South Africa and India can perhaps be seen as worst-case scenarios for efforts to mitigate the climate emergency—but are also, as the chapter has shown us, full of examples of environmentalism from below that *could* also serve as models globally. The two countries are far from alone. In late July of 2022, the Democratic Republic of Congo (DRC), home to one of the world's largest old-growth rainforests, auctioned off vast tracts of land to oil companies in a bid to become "the new destination for oil investments."[186] The sections of the rainforest where extraction is slated to take place feature tropical peatlands that store vast amounts of carbon, meaning that oil extraction there would be especially destructive. The auctions come only eight months after the central African country's president, Félix Tshisekedi, stood next to other political leaders at the United Nations climate summit in Glasgow and pledged not to allow drilling in the nation's rainforests for ten years. The pledge was in exchange for promises of $500 million from wealthy nations. But as wealthy nations in the Global North have themselves abandoned efforts to cut back on the use of fossil fuels after the Russian invasion of Ukraine sent oil and gas prices soaring, Tshisekedi has reneged on his pledge. The country's lead representative on climate issues, Tosi Mpanu Mpanu, said that the auctions will generate revenue for anti-poverty programs in Congo, which are of far more immediate importance than environmental issues: "Our priority is not to save the planet," Mpanu Mpanu commented.[187] Revealingly, the minister said nothing of the Indigenous Congolese people whose lands would be destroyed by oil exploration and extraction.

The move to open up Congo's forests to oil extraction, which will decimate not only the forests but also the wildlife and people who inhabit the forests, repeats a similar sorry history in Ecuador. In 2007, under President Rafael Correa, a trust fund was set up into which wealthy countries could deposit money to prevent the government from licensing oil exploration in the Yasuní National Park, one of the world's most biodiverse regions. Correa hoped to raise $3.6 billion, but, six years later, only $13 million had been deposited in the conservation fund. Correa denounced the lack of support for his Yasuní plan. It turned out, however, that his government had been secretly conducting negotiations with a Chinese bank to drill for oil in Yasuní even while Correa had been pledging to conserve the rainforest.[188]

The arguments that leaders in the DRC and Ecuador have made do raise issues of equity and extraction. In a paper exploring this topic, the researchers Greg Muttitt and Sivan Kartha suggest that if extraction-dependent poor countries are to be required to phase out extraction quickly for the global common good, the costs of this rapid transition should not be borne solely by poor countries alone.[189] Instead, Muttitt and Kartha argue, the largest burden should be shouldered by those with the greatest financial, technological, and institutional capacity, according to long-established UN Framework Convention principles of respective capacities—summarized in the maxim of "common but differentiated responsibilities," or CBDRs, codified at the 1992 Earth Summit in Rio.

This implies that extraction should be reduced fastest in the rich countries, where the social costs of doing so are the least, and that poor, extraction-dependent countries can reasonably demand compensation from the wealthy countries—which have, after all, gotten this way by colonizing the atmosphere with their carbon emissions, often derived from extraction of resources in Global South countries. Equity in extraction would imply not just phasing down extraction but also enabling a just transition for workers and communities, and curbing extraction in a manner consistent with environmental justice, so that extraction no longer violates people's rights and pollutes their environment.

These proposals for equitably ending extraction are admirable horizons of struggle, ones which focus the demands for climate reparations articulated by the global climate justice movement in the Cochabamba Declaration. Unfortunately, the fossil oligarchs are winning and the world is far from reversing ecological colonialism. Funding for just transition in the Global South has not been forthcoming. Two decades ago, at the UN Climate Summit in Copenhagen, rich nations promised to channel $100 billion a year to poorer nations by 2020 to help them adapt to the devastating impact of climate change and mitigate further rises in temperature. But, according to analysis by Oxfam, rich nations are falling $75 billion short of this goal.[190] In addition, according to Oxfam, 70 percent of public climate finance was given out as loans rather than grants, meaning that poorer countries are likely to be pushed deeper into the debt traps that have led to systematic dismantling of their public sectors during the decades since Global South debt ballooned in the 1980s.

This enduring refusal to decolonize financial structures and pay climate reparations makes a mockery of the pronouncements of institutions like the World Bank concerning just transition. In a 2021 report on transitioning away from coal, the World Bank lays out its ideas about policies "to support coal regions confronting the realities of decarbonization and help lay the groundwork for achieving a just transition for all."[191] The study offers reasonable-sounding policy suggestions, including the need to address both the informal and formal segments of coal workers through a combination of local and national programs. "Government's role in the transition process," the report states, "needs to be multi-faceted and proactive."[192]

But the World Bank continues to support neoliberal policy suggestions, such as the auction system for renewable energy, that undermine the public sector. Worse still, although it announced that it would cease all funding for fossil fuel projects in 2014, the World Bank is still bankrolling fossil fuels, hiding behind ideas of "energy access."[193] Money continues to flow to countries like India to build coal-fired power plants, either through arms of the World Bank like the International Finance Corporation (the entity that funded the seaside Tata Mundra plant); through funding the transmission lines that carry electricity produced in coal-fired power plants to cities; through "advice" to governments that often encourages them to grease the wheels for Big Oil and King Coal; or by lending to Wall Street firms, which then invest in fossil fuels. In 2020, for instance, 52 percent of IFC funding went to Wall Street.[194] India's big private coal company, Adani, has become the biggest issuer of offshore bonds in India, raising more than $9 billion from foreign investors in the past five years.[195]

Just transition will only be possible in poorer nations like India and South Africa when international financial institutions definitively cease all lending to fossil fuel–related projects, in all sectors of the industry, from extraction to transport to production and distribution. Climate justice activists in Global North and South countries can join in continuing to pressure for climate reparations as part of a Global Green New Deal. These funds should be devoted to a just transition away from fossil fuels and to renewable sources of energy that provide steady, well-paying jobs for workers.

The struggle to build such alternatives is already playing out alongside the fight to shut down polluting fossil capitalist projects that characterizes Global Blockadia. As we have seen, activists in Global South nations are fighting tenaciously to retain the community control over land that is central to Blockadia. This is a fight not just to challenge oppressive material conditions but to retain collective and culture.

As the activist Bhumika Muchhala argues, the colonial construction of humanity and the extractivism that it legitimates is predicated on two key falsehoods: that nature is dead and that land is empty.[196] If nature is dead, then it can be boundlessly exploited for its enriching resources. If the land is empty, then Indigenous and rural communities can be displaced or eliminated. According to Muchhala, forging a Global Green New Deal will hinge on establishing a "new social contract rooted in an ethical commitment to intersectional equity and justice," a contract that "is at the heart of a decolonial and feminist future." As we have seen from examples like the assassination of Mama Fikile in rural South Africa and the struggle of the Adivasi-led Chhattisgarh Bachao Andolan against mining in India, people whose land and lives are menaced by extraction are leading the fight for a just transition away from fossil fuels. At the core of the equity and justice that Muchhala imagines must be a fight against the colonial logic of extraction. Yet this fight involves not simply blocking more fossil infrastructure and extraction but also constructing democratically controlled forms of renewable energy as an alternative. We have a world to win, and it is grounded not just in a decolonial humanity but in public power.

# CHAPTER 4

# AGAINST FORTRESS CONSERVATION

Goanburah Kealing went out one morning to catch some cows that had wandered past the edge of his village. He never returned. Kealing, a young man of the Karbi tribe indigenous to the province of Assam in eastern India, was shot dead by guards from the Kaziranga National Park, on the edge of which his village is located. Guards in Kaziranga maintain a shoot-on-sight policy against local people, whom they accuse of seeking to poach the park's wildlife. The BBC reported in 2017 that over the two previous decades, 106 people had been killed by park guards; only one guard had died in encounters with purported poachers during that time.[1]

The park's anti-poaching policies have sparked repeated protests in Assam. For example, Pranab Doley and Soneswar Narah, activists with the Jeepal Krishak Shramik Sangha, a group that works for the rights of Indigenous people living in and around the Kaziranga National Park, were detained for weeks by the police after claiming publicly that shootings in the park amounted to a form of extrajudicial killing.[2] The Jeepal Krishak Shramik Sangha had staged a protest shortly before Doley and Narah's arrest in response to the announcement of plans to recruit ninety men to the Assam Forest Protection Force, a specialized armed force controlled by the national park's management.[3] In addition to decrying extrajudicial killings, Doley and Narah's group argued that park guards should be recruited from communities living close to the national park. The group also demanded compensation for community members killed by guards and by animals in the park.

Few tourists who visit the Kaziranga National Park are aware of these local protests against park policies. For these predominantly Western visitors, and for the mainstream wildlife conservation organizations who fund combat training for park guards as part of their efforts to stamp out poaching, the park is a roaring success. Kaziranga is blessed with abundant numbers of elephants and water buffalo. It has some of the highest tiger concentrations of any park in India. Most famous, though, is the park's population of Great Indian One-Horned Rhinos. On the brink of extinction a century ago, the rhino population in the park now stands at roughly 2,400. In the early 2010s, a spate of killings of these rhinos sparked alarm among park managers and political officials.

But efforts to conserve the park's population of rhinos are motivated not just by love of animals: tourist visits to Kaziranga bring in hundreds of thousands of dollars for the Indian government annually, and also constitute a major source of revenue for local hotels, restaurants, guides, and tourist agencies. The park is big business, and threats to its wildlife are taken very seriously by local politicians—as is underlined by the Assam government's recent announcement that it intends to deploy police commandos to the park to intercept poachers.[4]

Killings by park guards are not the only threat Kaziranga poses to local people. In 2020, the Assam government approved an expansion plan for the park. The announcement stated that the 3,000-hectare expansion would come from removal of "encroachments" on park land.[5] Expansion of the park and increasing militarization of its guards to deal with supposed "encroachments" spell loss of land, disruption of community, displacement, and, in some cases, deadly violence for the local people living on the outskirts of the park.

This was not the first time that the park has expanded. Kaziranga was established as a "reserve forest" in 1905 after Mary Curzon, wife of the British viceroy of India, Lord Curzon of Kedleston, failed to see any rhinos during a wildlife sightseeing visit to the area.[6] Lord Curzon set aside over 57,000 acres of land for the park, with more land added in the following three years.[7] After India's independence, even more local land was added to the park: indeed, Kaziranga doubled in size from the time when it was declared a national park in the mid-1970s to 1999. These expansions generated significant tensions with local communities, whose lands were confiscated to make way for the park's growth.[8]

Government officials tended to justify park expansion by disparaging the local population. A petition to the Assam High Court filed in 2012 by a member of the provincial legislative assembly representing the Hindu nationalist Bharatiya Janata Party (BJP) gives a sense of how government officials treated communities resisting expansion plans. In this petition, the BJP official describes local communities as encroachers and poachers, and argues that authorities should evict "illegal Bangladeshis" living on the park's periphery.[9] Although such blatant and patently false Islamophobia is atypical, state power has tended to support the dispossessions inherent in park expansion. In October 2015 the High Court ruled that the evictions could go forward, and one year later, two villages were cleared by armed forest guards. A BBC film crew, on location for the filming of a documentary about Kaziranga's policies, happened to catch the evictions on camera: police wielding long wooden cudgels known as *lathis* wade into a crowd of seated women protesters, beating them until they rise to their feet and run away.[10] In response to these attacks, BBC journalist Justin Rowlatt explains in the documentary, village men began throwing stones. Police reacted by opening fire on the villagers with live rounds, killing two people. After the villagers were dispersed, park guards rode elephants into the village and proceeded slowly and methodically to demolish every home.

The violence in and around Kaziranga National Park is symptomatic of the contemporary crisis of conservation. After all, despite the best efforts of well-funded international conservation organizations, we are living through an age of epic defaunation or animal slaughter.[11] In its *Living Planet Report 2020*, the World Wide Fund for Nature (WWF) reports a shocking 68 percent average decline in numbers of birds, amphibians, mammals, fish, and reptiles just since 1970.[12] This "sixth extinction wave" is characterized by what scientists term "extirpation," or the extinction of both specific animal species and entire populations on a local or regional scale.[13] This biodiversity crisis is affecting not just animals but also other forms of life, including insects, plants, and fungi. A 2020 report from the Royal Botanic Gardens, Kew, for instance, stated that some 40 percent of the world's planet species are now threatened with extinction.[14] And it is not just specific species of animals, plants, and fungi that are being annihilated: we are also witnessing dramatic declines in the abundance of local species across all taxonomic groups—although some groups and geographic regions are more intensely affected than others. The scale of annihilation of the planet's life forms is so immense that it has become a primary driver of global environmental change in its own right.[15]

The dominant tendency has been to blame an undifferentiated humanity for these shocking numbers of extinction and biodiversity loss. As many critics have argued, this is the import of the now well-known term "the Anthropocene," which attributes the massive environmental changes undergone by the planet in recent centuries to a homogeneous humanity. The term conveniently obscures the fact that only a small segment of humanity—hyper-consumptive global elites—is responsible for the vast majority of environmental impacts, and that a specific economic and political system—capitalism—is driving this wave of annihilation.[16]

A similar obfuscating lens has been applied to the biodiversity crisis. For example, in her best-selling book *The Sixth Extinction*, journalist Elizabeth Kolbert seems to attribute extinction to writing, or, more generally, to the human capacity to represent things using language. It is this representational capacity, she argues, that is the root of our unique ability as a species to reorder the world. Kolbert thereby sets up a firm distinction between people and nature, a division that is fundamental to dominant forms of conservation, which hinge on walling nature off from people. In addition, although Kolbert points the finger of blame at people reading her book, before doing so she names the usual suspects—poachers in Africa and loggers in the Amazon—without providing any context for why people in these places might engage in such behavior. Her suggestion that small-scale hunters and subsistence farmers are responsible for the Sixth Extinction is a bit of disproven Western ideology that should be seen as thoroughly racist. Given the violence meted out to people represented as poachers in parks around the world, Kolbert's accusation tacitly legitimates the policies carried out in places like Kaziranga. This kind of ahistorical analysis places a bull's-eye on the backs of many people in the Global South.

Spurred by this idea that people in general must be kept separate from wildlife if biodiversity is to be preserved, big conservation organizations are ramping up what critics call "fortress conservation." This long-dominant approach to conservation involves setting up increasing numbers of national parks, wilderness areas, nature reserves, and similar enclosed territories, all of which are examples of what conservation organizations call "protected areas" (PAs). With the intensification of the extinction crisis, the conservation industry has thrown into overdrive its efforts to enclose more land. As historian Mark Dowie reports, there were only 1,000 official PAs in 1962, when many of today's big conservation organizations did not yet exist. Today there are 108,000 PAs around the world. According to Dowie, "the total area of land now under conservation protection worldwide has doubled since 1990, when the World Parks Commission set a goal of protecting 10 percent of the planet's surface."[17] This goal has been dramatically surpassed: according to the *Protected Planet Report 2020*, at least 17 percent of terrestrial and inland water, and 10 percent of coastal and marine areas, are conserved.[18] Since 2010 alone, PAs covering almost 21 million square kilometers have been added. If this territory were contiguous, it would be larger than the Russian Federation—a stunning amount of land.

The plan is to roughly double this enclosed land. In 2016, renowned biologist E. O. Wilson published *Half-Earth: Our Planet's Fight for Life*, a headline-grabbing manifesto for expanding PAs around the world. The following year, influenced by Wilson's arguments, a group of scientists published a paper laying out a plan for what they called a "Global Deal for Nature" as a companion to the Paris Agreement.[19] They argued that their plan would "promote increased habitat protection and restoration, national- and ecoregional-scale conservation strategies, and the empowerment of indigenous peoples to protect their sovereign lands."[20] The plan suggested that in order to maintain a livable planet, governments would need to protect 30 percent of Earth's land and sea and sustainably manage another 20 percent. For the big conservation organizations that quickly endorsed this plan, the massive expansion of PAs called for in the Global Deal for Nature would be a great victory. For the people whose land would be taken to create and expand PAs, and who are threatened with torture and death if they venture into lands managed sustainably by previous generations, conservation today is less of an unblemished boon. Indeed, some maintain that it is a new form of colonialism.

As the conservation industry encloses more and more of the lands collectively owned by Indigenous people, peasants, and pastoralists around the world, the dispossessed are increasingly rising up in opposition. Their cry: Decolonize conservation! For the growing movement opposing what they call the "conservation industry," the best way to fight against biodiversity loss and climate change is to respect the land rights of Indigenous peoples and other local communities, who protect 80 percent of the world's biodiversity. As expressed in the *People's Manifesto for the Future of Conservation*, a document that emerged from a counter-summit of groups challenging the conservation industry, this means that there must be an end to the dominant model of conservation, which "is often violent, colonialist, and racist in approach—seizing and militarizing the land, criminalizing and destroying the ways of life of Indigenous and local communities, while ignoring their knowledge."[21]

Given the scale of the biodiversity crisis, we certainly need forms of conservation—just not the exclusionary models of colonial conservation that dominate the movement today. The *People's Manifesto* calls for an alternative mode of conservation grounded in the work of Indigenous people and local communities to preserve their lifeworlds. It also argues that industrialized nations must agree to significant reductions in the activities that produce greenhouse gases, rather than simply trying to shift extraction of resources to more vulnerable countries, and must develop policies to enforce these shifts on corporations headquartered in the core imperial nations. As the *People's Manifesto* puts it, "We need a conservation model that fights against the real causes of environmental destruction and is prepared to tackle those most responsible: overconsumption and exploitation of resources led by the Global North and its corporations."[22]

This chapter explores the context for this campaign to decolonize conservation. As the history of fortress conservation policies makes clear, it has its roots in European powers' recognition of the destruction of ecosystems in colonial territories such as the Caribbean and India. Conservation policies in such territories and in settler colonies like South Africa and the United States were often animated by a desire to protect valuable timber supplies as well as "game," the charismatic megafauna that wealthy white men like to hunt. In recent decades, big conservation NGOs have put forward a slate of "Nature-based Solutions," or NbS, market-oriented approaches that rely on poor countries to absorb the carbon emissions generated by global elites. The plan to expand protected areas to cover 30 percent of the earth's terrestrial surface by 2030—promoted with the catchy tagline "30x30"—is only the most prominent of the Nature-based Solutions currently being promoted by the conservation industry. But, as the *People's Manifesto* makes clear, we can learn from the robust alternatives proposed by the movement to decolonize conservation.

As the intersecting climate and extinction crises generate increasingly extreme environmental conditions, the pressure on political and economic elites to do something about the crisis is only going to intensify. As the example of conservation shows, it is by no means certain that such climate action will be socially just or that it will lay the foundations for a sustainable future. If conservation may be seen as one of the world's oldest and most materially significant forms of environmentalism, it is also a key site for efforts to win emancipatory, decolonial futures.

## Colonial Conservation

When a government and conservation organizations declare a protected area, what does it entail? Does it take the form of a "reserve" like Kaziranga, where local people are banned from entering the sequestered land? Mainstream conservation organizations tend to ignore or downplay this question, with the result that the declaration of PAs may be seen as nothing but another land grab, one whose scope is truly breathtaking. As the World Rainforest Movement puts it, "There has been no real recognition of the underlying colonial and racist roots upon which the dominant conservation model is founded. As a result, the management of Protected Areas continues to be linked—directly or indirectly—to forced evictions, harassment, violence, and sexual abuse of women and children, human rights violations, deforestation and militarization of forest peoples' territories."[23]

What, precisely, are the "underlying colonial and racist roots" of the dominant conservation model? For the historian Richard Grove, the origins of conservation and Euro-American environmentalism go back to the ecological destruction wrought by racial capitalism and colonialism.[24] Grove's research, importantly, challenges standard accounts of environmental thinking, which trace the movement's origins back to the writing and activism of US-based figures like Henry David Thoreau and John Muir in the late nineteenth and early twentieth centuries. For Grove, environmental thought and conservation policies were, instead, a reaction to the alarming ecological shifts provoked by the intensive exploitation of the land that characterized colonial European plantation economies on islands in the Caribbean and Indian Ocean. Plantations involved the systematic destruction of indigenous plant life on islands colonized by Europeans, part of what anthropologist Sidney Mintz called "a synthesis of field and factory." The cultivation of monocultural crops by forced labor (indentured servitude first, followed by slavery) rapidly depleted soil nutrients, and the result was devastating in both human and ecological terms.[25]

This conjunction of relentless racial and ecological violence in the pursuit of early forms of capital accumulation produced unnerving changes in island ecologies. Science studies scholar Donna Haraway and anthropologist Anna Tsing call this process the Plantationocene—the sweeping environmental changes brought about around the world by the expansion of the European colonial plantation economy throughout the tropics after 1500.[26] The term is a stinging rebuke to the idea of the Anthropocene, a geological term taken up in popular discourse that evinces the same sweeping generalization of agency and responsibility as we have seen characterizing dominant interpretations of the Sixth Extinction.

The changes signified by the term Plantationocene were noted by a nascent network of colonial botanists as early as the mid-eighteenth century. Grove documents how these scientists observed the alarming impact of deforestation and degradation on small tropical islands such as St. Vincent in the Caribbean, leading to the emergence of a theory of what was called "dessicationism," which connected the destruction of forests to changing patterns of rainfall and regional aridification. But colonial botanists were not content with simply observing and analyzing the problem: they also planned solutions. The strategies they proposed were embodied in what Richard Grove characterizes as the first piece of modern European environmental legislation, the Kings Hill Forest Act.[27] Passed on the Caribbean island of St. Vincent in 1791, the act arrogated to the state the right to "appropriate for the benefit of the neighborhood the Hill," with the goal of "enclosing the same and preserving the timber and other trees growing thereon in order to attract rain."[28] The Kings Hill Forest Act was thus significant not simply for recognizing the link between deforestation and environmental degradation, but also for proposing both conservation of existing forests and programs of afforestation as solutions to environmental crisis. The ideas for conservation pioneered on colonized islands like St. Vincent were to have dramatic, world-changing effects.

Conservation was also becoming an issue at this time in the vast forests ruled over by the British East India Company (EIC) in India. Keen to increase agricultural productivity in India to generate more tax revenue, the EIC oversaw policies of forest clearance in order to open more land for cultivation. The growth of British rule in the eighteenth century thus led to a rapid increase in the rate of forest destruction.[29] Nonetheless, concern about the rapid depletion of India's forests grew quickly after the loss of the American colonies and the initiation of wars against the French Revolution and Napoleon.[30] In this context of heightened inter-imperial competition, timber reserves became a resource of primary strategic concern, particularly for the Royal Navy. In 1805, the EIC therefore set up a forest committee to investigate the possibility of using mature teak trees from India's forests.[31] The findings of this inquiry were devastating: the committee flatly contradicted previous assumptions about the unlimited nature of teak supplies in India's forests, documenting instead the virtually complete exhaustion of wood in easily accessible regions of the subcontinent and the logistical and economic difficulties of tapping distant, unmapped forests.[32] The committee recommended that India's teak forests be protected, and that timber be harvested based on the principles of continual replenishment of supplies that characterized "scientific" forestry in continental Europe. A general proclamation was issued prohibiting all unauthorized felling of teak trees, and unclaimed lands were declared property of the crown. In 1806, the directors of the EIC appointed a po-lice officer named Captain Watson as India's first conservator of forests. As Gregory Barton observes in his history of empire forestry, this was a watershed moment: "For the first time a modern nation-state appointed an officer with vague police powers over 'waste' lands unclaimed by private owners, and charged him with the preservation of forests."[33]

The growth of British colonial power in India from the early to mid-nineteenth century nonetheless ensured continued overexploitation and depletion of the subcontinent's forests. For example, Britain's con-struction of a railway system throughout India required massive supplies of timber for rail track supports known as "sleepers." As a result, EIC administrators' concern about the depletion of India's forests continued to grow. These worries only increased as imperial scientific networks spread word of the sensitivity of Caribbean island ecosystems to environmental degradation. This mounting concern led to the formation of the Indian Forestry Department in 1864, and the passage of the Indian Forest Act the following year. The Forest Act of 1865 was the first India-wide forest legislation promulgated by the British, and, equally if not more important-ly, the first broad-based environmental law of the nineteenth century.[34] The act empowered the colonial government to declare any land covered with trees as government forests and to issue rules for conserving them. Removal of all forest produce was prohibited, and local governments were authorized to arrest violators.

The magnitude and violence of this sweeping legal enclosure of India's forests is evident from the resistance it generated even among certain co-lonial functionaries. Most exemplary is the opposition to the act voiced by members of the Board of Revenue of Madras, who wrote:

> All the jungles and forests of this Presidency are within village boundaries, and the people residing in or near them have, from time immemorial, had the right to take leaves for manure, fire-wood for their own use, and timber for agricultural purposes, to graze their cattle at certain periods. These rights have been re-peatedly recognized by Government, and are now scrupulously respected. When, therefore, these and other similar existing priv-

ileges, as well as the rights of way which necessarily exist through forest tracts are taken into consideration, the operation of the Act under this Presidency will be very limited, and every prosecution under it will be met with the allegation (which the Forest Officers must disprove) that a right previously existed which vitiates the application of the Act.[35]

The colonial revenue collectors of Madras recognized the customary traditions and rights that had long permitted Indian village communities to engage in subsistence harvesting of India's forests. This meant that the forests belonged to those who lived within and alongside them, leaving virtually no forest land for the colonial government to claim. In a classic instance of hearing the voice of the subaltern through the statements of colonial officials, the revenue collectors worried that efforts on the part of the colonial government to prohibit forest communities from their customary practices would be seen as an effort to annul long-recognized rights. Indeed, the officials argued that "it is impossible to introduce the strict system of conservancy practicable in countries like Burma without exciting much popular discontent, and incurring serious risk of oppression."[36]

The Madras officials' concerns did not, however, shift the direction of colonial conservation policy. Indeed, the enclosure of India's forests gained speed in the period after the creation of the Forestry Department. In 1864, Dietrich Brandis was made inspector general of the Imperial Forest Service in India. Like many other forestry officials in British India, Brandis was German. The British Raj was forced to recruit German foresters since the replacement of wood fuel with coal in England after the eighteenth century meant that there was scant scientific attention to forestry in the imperial homeland. The German Forest Service was considered the most scientifically advanced in Europe, but it had also made its reputation by successfully prohibiting customary woodland uses by the Rhineish peasantry, and by suppressing an 1848 peasant uprising against forest enclosures in that region.[37] The young Karl Marx, who in the early 1840s was working as editor of the *Rheinische Zeitung*, wrote scathingly of the Prussian state's criminalization of the traditional practices of the forest-dwelling peasantry: "All the organs of the state become ears, eyes, arms, legs, and means by which the interest of the forest owner hears, sees, appraises, protects, grasps, and runs."[38]

Marx's critique was motivated by the expansion of state ownership and control over the forests in the Rhineland, measures that included the development of an apparatus for the scientific study and management of timber.[39] In 1837 the government passed the Forestal Theft Act, the legal modification that drew Marx's acerbic critique, and moved to establish a forest police to enforce this new legal regime. German peasants found that the usufruct rights that they had over the forest commons were suddenly revoked: written permits were required, for example, to gather berries and mushrooms in the woods. Seventy-two percent of criminal prosecutions in the region during the 1830s were for forest offenses. So rankling was this state control of the forest commons that it contributed in no small part to the revolutionary uprisings of 1848: as Peter Linebaugh puts it, "The great rural jacqueries of March [1848] that swept southwestern Germany were in part united by their common attempts to reappropriate the wealth of the forests."[40]

It was these experiences of "scientific management" and draconian policing that made the reputation of German silviculture experts like Dietrich Brandis. After his appointment as inspector general of the Imperial Forest Service, Brandis set out on a series of fact-finding tours of India's forests, during which he claimed to have discovered endemic forest "abuses." The destructive activity that he documented was not connected to the commercial exploitation of forests that previous inquiries had concluded were depleting India's forests at alarming rates. Instead, Brandis argued that the destruction of India's forests was the fault of the country's "tribal populations." In a memorandum he published that galvanized support for the passage of the 1865 Forest Act and its more draconian successor act of 1878, Brandis wrote:

> If we take a review of the present impoverished state of a large proportion of the forests in British India, we come to the conclusion that certain main causes have co-operated to reduce them to the present unproductive condition. Indiscriminate felling and reckless mutilation of trees, the lopping of branches for fodder

and litter, the wholesale cutting of young trees for fencing and roofing, have contributed much . . . to prevent the improvement of the Forests, and many tracts have been entirely denuded.[41]

Brandis essentially blamed customary users of India's forests, including hunters and gatherers, subsistence cultivators, and villagers living nearby.[42] His goal was to render the forests economically productive for the empire, hence his account of their "present unproductive condition" and the destructive practices that prevent their "improvement." This latter keyword is particularly important, for by the time Brandis was writing it had become central to ideological justifications for enclosure of commonly held lands and dispossession of their inhabitants in both England and its colonies. To understand the full import of Brandis's argument, and its subsequent and enduring impact, we need to take a brief detour back to the writings of John Locke.

Locke is often remembered as one of the early philosophers of liberalism and empiricism, yet he also contributed key arguments to theories justifying the enclosure and privatization of land. In his *Two Treatises on Government*, Locke argued that all people are equal in their natural state, and that humanity therefore originally held all land in common. Private property developed when people began working the land to grow crops, in the process making land more productive. By coaxing the land to bear fruit through the application of their own labor, Locke argued, a person gained ownership of the land. Private property, in other words, derives from the labor of "improvement" carried out by hard-working individuals. In this way, "wasteland" was transformed into fruitful property. But it was not just purportedly vacant land that could be thus appropriated: for Locke, land held in common by subsistence-based communities was also wasted inasmuch as it was not being used to its fullest potential. Locke reasoned that landowners intent on developing common land had a right to dispossess the inhabitants and enclose such land, since the increased productivity their possession of the land would generate stood to benefit humanity in general in the long run. Locke thus wrote that "he who appropriates land to himself by his labor does not lessen but increase the common stock of mankind."[43]

Locke's arguments offered a potent justification not just for the many acts of enclosure that dispossessed rural peasants of their land across Britain in the eighteenth and nineteenth centuries, but also for settler colonialism. Indeed, as a number of postcolonial discussions of his work have pointed out, Locke's labor theory of property was essential to the ideological legitimation of the conquest of the Americas.[44] He wrote of the "wild woods" and "uncultivated waste of America, left to nature, without any improvement, tillage, or husbandry."[45] The implication was clear: by stripping Native Americans of their land, settlers were fulfilling God's will that land should be transformed from its conditions of uncultivated waste and rendered fruitful through hard labor. In this way, property was established through a continental-scale act of theft. It should come as little surprise that Locke was an owner of plantations in both Ireland and the British colonies in North America.

Although Locke clearly wrote in order to justify colonization of the Americas, his labor theory of property also played an important role in the British conquest of India.[46] As geographer Vinay Gidwani has argued, EIC officials used Locke's categories of "waste" and "wasteland" to suggest that India's Mughal rulers were illegitimate as a result of their failure to improve the country by rendering its farmland sufficiently productive.[47] EIC bureaucrats' descriptions of the desolate condition of portions of India were used to legitimate a discourse of British cultural superiority over the subcontinent's Mughal rulers.[48] Such observations of the denuded state of hills and countryside in Muslim-ruled regions of India supported the notion that European colonial rule would confer environmental benefits on the country by ending such depleting practices. This stance as an improving landlord came to characterize British rule in India more generally.[49] The allegation of wastefulness also offered a perfect justification for stripping rights from the inhabitants of India's forests. By describing the supposedly wasteful practices of India's forest-dependent peoples, imperial foresters such as Dietrich Brandis implicitly invoked Locke's rationale for colonial dispossession. Rather than seeing the practices of India's traditional forest-dependent communities for what they were—locally based systems of forest conservation and management that had been practiced sustainably for thousands of years—British accounts of the wasting of Indian forests laid the foundation for their co-optation by the state.

It was not simply that forest dwellers were allegedly depleting India's valuable forests by practices such as those Brandis itemized in his memorandum. In addition, they were an obstacle to the "improvement" of these forests—their scientifically managed transformation into resources that would generate capital for the colonial power. In fact, the British were intent on bringing about an ecological transformation of India's forests, replacing the more heterogeneous local varieties of trees upon which Indian forest dwellers and farmers had long depended with commercially valuable trees such as teak, sal, and deodar.

The allegation of wastefulness was instrumental to this transformation. Brandis and other imperial foresters conveniently ignored the excessive commercial exploitation of Indian forests under EIC rule during the late eighteenth and early nineteenth century, and instead held customary users of the forest commons responsible for the depleted state of these forests. We now know that even the worst alleged "abuses" engaged in by so-called tribal populations, slash-and-burn agriculture, are in fact highly efficient and ecologically safe when practiced under conditions of stable population growth.[50] Adivasi communities practicing what was known as *jhum*, or shifting agriculture, employed an extended fallow system to allow forest tracts to regenerate.

Nevertheless, Brandis's damning account of the practices of forest-dependent communities became the foundation for subsequent imperial forest law, which simultaneously dispossessed India's forest-dwelling people and criminalized their traditional customary practices. As the anthropologist Judith Whitehead writes, "The India Forest Act of 1865 carved out the categories of state forests and District Forests from the category of wastelands and defined productive versus wasteful uses (and users) of the forests. . . . Criticism of customary users of the forests as wasteful and destructive became a normative call to remedial action through enclosure, prohibition, and restrictions on customary practices by the newly formed Forest Service."[51]

The forest laws promulgated in British India became the model for laws and policies that enclosed previously common lands and community-stewarded natural resources across Asia, Africa, and the Americas.[52] Understanding conservation's long colonial history is an important counter to narratives that trace the creation of exclusionary PAs back to the creation of Yellowstone and Yosemite National Parks in the United States (in 1872 and 1890, respectively). Established in 1864 following a bloody war of extermination against the Miwok people, Yosemite in California was the first national park—yet the idea of setting aside certain areas of land to protect natural species that live there did not originate in the US. Indeed, while the history of exclusionary conservation in the US helped establish global paradigms whose negative influence remains to this day, as the history of the Forest Service in India shows, such practices were not exceptional and not even original to the US.[53] The violent expulsion of Native Americans who lived in and depended on the natural resources in the areas that became the country's famous parks is part of a broader—and necessarily violent—dispossession of Indigenous and forest-dependent peoples around the globe.[54] Furthermore, this massive global land grab was also tied structurally to the forms of enclosure that played out in the farmlands and forests of Europe during the age of colonial expansion.

Exclusionary forest laws didn't simply ban people from gathering vegetable matter in the woods; they also proscribed the hunting of animals. The word *poaching* is derived from a Middle English word referring to the bag or pouch in which peasants would hide animals they'd killed on land where hunting was reserved for the Anglo-Norman aristocracy. A poor person caught poaching in medieval England was punished by death by hanging. In keeping with these English traditions, the new forest laws in India restricted not just the gathering of forest produce but also small-scale hunting by forest dwellers.[55] At the same time, these laws simultaneously facilitated more organized *shikar*, or big game hunting expeditions, by British colonial officials. The result was an epic level of destruction of tigers, elephants, and other large animals, as British colonials sought to ape the bloodthirsty behavior of European aristocrats. The historians Madhav Gadgil and Ramachandra Guha write of the massacre of animals conducted in tandem with forest conservation:

Much of this shooting was motivated by the desire for large "bags." While one British planter in the Nilgiris killed 400 elephants in the eighteen sixties, successive viceroys were invited to shoots in which several thousand birds were shot in a single day in attempts to claim the "world record."[56]

After a half-century of this unbridled carnage, British hunters began to worry that many species of animals were being pushed to the brink of extinction. In 1899 officials convened the first International Conference dedicated to wildlife conservation measures, with officials from the major European colonial powers (Britain, Germany, France, Italy, Portugal, and Belgium) meeting in London.[57] The outcome was an agreement among officials to publish a list of species in danger of extinction, which were to be absolutely protected at all times—a forerunner to the Red List of Threatened Species, compiled by the International Union for Conservation of Nature.[58] In his account of colonial conservation efforts, Edward Buxton, an important member of the Society for the Preservation of the Fauna of the Empire, an organization founded in 1903 that is often referred to as the first international conservation organization, wrote that the killing of animals should be limited by the issuing of licenses that would stipulate a specific number of animals of each species that could be shot.[59] Buxton also argued that reserves or sanctuaries should be created throughout British colonial African territories in order to preserve various species.

African people are almost entirely absent from Buxton's discussions of wildlife conservation measures, although they do feature in the photographs that illustrate his book, posed alongside trophies of the hunt as symbols of servility to European manly vigor. Nonetheless, near the conclusion of his book Buxton does argue that Africans should be exempted from the prohibitions on hunting that he has outlined, saying that "it is neither possible or just to stop their hunting so long as it is confined to their primitive weapons, the poisoned arrow and spear."[60] He notes that unlike the "superfluous massacre" carried out by white men in Africa, "from time immemorial the destruction caused by the indigenous inhabitants has not appreciably diminished the stock." In addition to this backhanded acknowledgment of the sustainable character of African people's relations with wildlife, Buxton echoes the comments of the Madras revenue collectors concerning the dangerous effect of the Indian Forest Act, noting that a prohibition on Africans' hunting rights would likely spark resistance: "The land and the animals upon it are their birthright, and to interfere with it would surely cause trouble."[61] It would not be long before movements for African independence would underline the prescience of Buxton's cautionary remarks.

## Conservation after Colonialism

During the postcolonial era, Indigenous people, peasants, and other rural people across Africa and Asia continued to see their land dispossessed through reserves, game laws, and conservation policies. While Gadgil and Guha praise India's "massive network of parks and sanctuaries," which was largely constructed after independence in 1947, as a "magnificent achievement" that stands out from the overall ecological decline of the subcontinent, they nevertheless note, "In displacing villagers without proper rehabilitation, prohibiting traditional hunting and gathering, and exposing villages on the periphery to the threat of crop damage, cattle lifting, and manslaughter, the parks are, as they stand, inimical to the interests of the poorer sections of agrarian society."[62] Preserving biological diversity may be important, but Gadgil and Guha underline the devastating impact of conservation policies based on the precepts of the Western wilderness movement, including the belief that *all* human intervention is bad for the retention of biodiversity (i.e., that "wilderness" is a human-free zone), and that wilderness areas should be as large as possible. This has resulted, in India as in other parts of the Global South, in the construction of massive sanctuaries along the lines of Kaziranga National Park, in which there is "a total ban on human ingress in 'core' areas of national parks."[63] As we have seen, "human" in this instance refers only to local or Indigenous inhabitants, since in many if not most cases, one of the main reasons for these parks' existence is to attract the patronage of wealthy Western tourists. In this way, the paradigm of "fortress conservation" that constitutively excludes Indigenous people and forest-dependent communities continues to spread around the world.

Powerful mobilizations among Indigenous peoples in the 1970s and 1980s challenged the exclusionary foundations of conservation. With the independence of many so-called Third World nations in the post-1945 era, many Indigenous people found themselves locked into nations that did not recognize their rights. They effectively constituted a Fourth World, extending from Alaska to Tierra Del Fuego in the Americas, and including the Ainu of Japan, the Aborigines of Australia, the Maori of New Zealand, the Sámi of Scandinavia, and the tribal peoples of the former Soviet Union, China, India, Southeast Asia, and Africa. Under the pretext of conserving nature, postcolonial states continued colonial policies of expelling Indigenous people from their land. In the late 1960s, for example, up to six thousand Indigenous Batwas (Forest People) in the eastern Democratic Republic of Congo were expelled from their ancestral land in the freshly designated Kahuzi-Biega National Park, which would become a World Heritage site in 1980. The Batwa people sought justice through DRC courts without success, but in 1975 the Kinshasa Resolution adopted by the IUCN and the World Parks Congress recognized the importance of Indigenous traditional lifestyles in conservation and encouraged governments to provide for Indigenous peoples to turn their lands into protected areas without surrendering ownership, use, and tenure rights.[64]

In the wake of the dispossession carried out by European colonizers and continued by postcolonial states, the Kinshasa Resolution marked the struggle among Indigenous groups around the world for recognition of their customary rights to land. Some Global South nations initiated reforms to recognize these rights legally, which also necessarily involved rethinking fortress conservation.[65] For example, Brazil's post-authoritarian constitution of 1988 recognized the rights of Indigenous peoples to their traditional lands, which opened the way for formal titling of large areas of the Brazilian Amazon to customary rights-holders. On an international level, political campaigns led to the recognition of customary rights of Indigenous peoples to lands, territories, and resources, as well as to the restitution of lands taken from them without prior, informed consent. The struggle for the recognition of these rights featured prominently in the negotiations that led to the United Nations Declaration on the Rights of Indigenous Peoples (UNDRIP), which was formally adopted in 2007.

Article 19 of UNDRIP specifies that "states shall consult and cooperate in good faith with the indigenous peoples concerned through their own representative institutions in order to obtain their free, prior and informed consent before adopting and implementing legislative or administrative measures that may affect them."[66] While international law has progressively advanced toward recognizing this right, Indigenous peoples' customary collective right to land is not recognized in many individual states' domestic law, leaving their land still regarded as vacant—an extension of colonial-era *terra nullius* doctrine.

Studies produced in this period revealed that 50 percent or more of protected areas had been set up on land traditionally used or occupied by Indigenous peoples (in the Americas, of course, all PAs are on Indigenous land). At a United Nations meeting during the UNDRIP negotiations in 2004, an Indigenous delegate declared that conservation was the biggest enemy of Indigenous peoples.[67] That same year, delegates at an Indigenous mapping conference signed a declaration stating that "conservation has become the number one threat to Indigenous territories."[68] This political pressure led conservation organizations to recognize the importance of what was referred to as "community-based natural resource management," or CBNRM, by Indigenous peoples and other local communities. The outcome of this shift included community-based wildlife management programs such as Zimbabwe's CAMPFIRE program, community forestry programs in countries like Mexico and Nepal, and locally managed marine areas in the South Pacific. As the first community-based wildlife conservation project, the CAMPFIRE program is particularly significant. By establishing quotas that permit visiting hunters to kill a limited number of animals for significant fees, the program benefits local schools, clinics, and wildlife research facilities. Through this program of hunting permits that generate tangible revenue, the program aims to convince local populations that wildlife is an economic asset rather than simply an impediment to farming. A recent assessment of CAMPFIRE by a pair of scholars based in South Africa found that local people acknowledged the beneficial impact of CAMPFIRE on infrastructure. But they nonetheless did not feel that it had successfully reduced overall poverty in the area.[69]

The fight of Indigenous people and land-dependent communities for recognition and rights led prominent conservation organizations to recognize Indigenous rights in documents such as the Durban Action Plan, an outcome of the 2003 IUCN World Parks Congress, which called for Indigenous representation in PA management and restitution of PA land taken from Indigenous people without their free and informed consent. In 2010, the UN Convention on Biological Diversity adopted the Aichi Targets on Biodiversity, which describe both PAs and "other effective area-based conservation measures," or OECMs, as ways to safeguard ecosystems, with the latter category potentially including Indigenous and community-managed lands. Such changes in the language of the world's most powerful conservation organizations are undeniably hard-won victories. Nonetheless, most reforms of protected-area legislation since the 2003 World Parks Congress are focused on enabling co-management of reserves, or on making provisions for communities that already own land to include their territory in national PA systems. Remarkably, there is not a single country where PA laws recognize community land ownership.[70] Meanwhile, states continue to arrogate the right to land that they won during the colonial era. This material fact vitiates much of the rhetorical respect for the rights of Indigenous peoples found in recently produced documents. As an exhaustive assessment of Indigenous peoples and local community land rights by the nonprofit Rights and Resources Initiative concludes, "Although some progress has been made in the last decade, national laws still fall far short of guaranteeing respect for customary rights in protected areas."[71]

The problems with dominant forms of conservation are reflected in the language used by some of the world's most powerful and influential organizations, including the International Union for Conservation of Nature. Founded in 1948, the IUCN is a network of more than one thousand environmental, scientific, and governmental organizations that compiles and publishes the well-known Red List of Threatened Species. The importance of the IUCN in the world of conservation can hardly be overstated: it has consultative status with the United Nations and strong links to many governmental and nongovernmental organizations, as well as—controversially—many businesses. Indeed, as Bram Büscher and

Robert Fletcher explain in their book *The Conservation Revolution*, the IUCN exemplifies what has become known as "corporate conservation."[72] Büscher and Fletcher describe the IUCN as "an increasingly dense and self-referential network" that binds together three key sets of powerful interests: big non-governmental organizations (BINGOS) like Conservation International, The Nature Conservancy, World Wide Fund for Nature, and the Wildlife Conservation Society; intergovernmental financial institutions like the World Bank and the International Monetary Fund that have been key to promoting neoliberal paradigms around the world over the last half century; and prominent business "partners" coordinated within the World Business Council for Sustainable Development.[73]

The existence of this powerful network is a problem since it means that nominally conservation-oriented organizations like the WWF collaborate with some of the most environmentally destructive capitalist corporations in the world. Indeed, they actually share many of the same staff.[74] In Büscher and Fletcher's words, conservation has become "more intensely and overtly capitalist in its goals, expressions, imaginations and ways of operating."[75] And, conversely, the IUCN's incestuous mingling of powerful business interests and conservation organizations also means that conservation has become increasingly important to global capitalism—a fact evident from the nearly ubiquitous references to sustainability in contemporary corporate advertising.

In 2021 the IUCN network held the World Conservation Congress in Marseille, during which it generated a document called *The Marseille Manifesto* that embodies the priorities of the organization and its many affiliates.[76] At the outset, the manifesto recognizes that "the climate and biodiversity emergencies are not distinct, but two aspects of one crisis. Unsustainable human activity continues to compound the situation, and threatens not only our own survival but the foundation of life on Earth."[77] Although the IUCN's recognition of the interrelated character of the climate and biodiversity emergencies is a very positive thing, the manifesto reproduces exactly the same generalizing interpretation of the factors driving the emergency that, as I have noted, characterizes the idea of "the Anthropocene."

Although the manifesto offers no direct mention of the role of racial capitalism in environmental crises past and present, it does acknowledge that "economic 'success' can no longer come at nature's expense. We urgently need systemic reform."[78] While this statement seems at first glance to admit that the dysfunctional global capitalist system is responsible for ecocide, closer scrutiny reveals the same kind of generalization and equivocation that characterizes liberal discourse about the climate crisis in general. Precisely what does the scare-quoted reference to economic "success" mean? Whose success is being described, and who is the audience for this statement? Orthodox economists? Corporate CEOs of the transnational capitalist enterprises with which the IUCN partners? The general public in core imperialist nations? If the target of criticism is purposely left vague, so is the mode of necessary transformation: the manifesto acknowledges that the problem is systemic, but in the same breath suggests that this system needs "reform" rather than transformation. In other words, the economic and political system that is generating environmental crises can be left in place, and all that is necessary are tweaks in how this system works. While acknowledging that the planet is on the brink of catastrophic environmental breakdown, the manifesto also reassures readers that this extreme situation does not warrant any revolutionary transformations of business as usual.

*The Marseille Manifesto* does get specific in one particular area: its urgent call for the expansion of protected areas around the globe. The document states that "the [World Conservation] Congress implores governments to set ambitious protected area and other effective area-based conservation measure (OECM) targets by calling for at least 30% of the planet to be protected by 2030."[79] The IUCN here clearly adopts the media-friendly "30x30" slogan, a demand that was also taken up by the working group that drafted a post-2020 global biodiversity framework for the Convention on Biological Diversity (CBD). It is worth mentioning that the manifesto does register the potential dangers of such an expansion of PAs. In fact, in the following sentence it states that "these targets must be based upon the latest science, and reinforce rights—including free, prior, and informed Consent—as set out in the UN Declaration on the Rights of Indigenous Peoples. IUCN must boost the agency of indigenous peoples and local communities, and reduce biodiversity loss at scale."[80]

It is not particularly surprising that the IUCN is making these rhetorical gestures toward inclusion, democracy, and the empowerment of

Indigenous people. Adoption of the 30x30 goal is likely to have a devastating impact on Indigenous people and other land-dependent communities. A scientific analysis of the impact of E. O. Wilson's Half-Earth proposal published in the journal *Nature Sustainability* in 2019 concluded that "at least one billion people live in places that would be protected if the Half Earth proposal were implemented within all regions."[81] A subsequent analysis by Rainforest Foundation UK using this same data concluded that meeting the 30x30 target accepted by the Conference of Parties to the CBD "could directly displace and dispossess 300 million people, with many more being indirectly affected."[82] The authors of the *Nature Sustainability* article note that "implementing Half Earth [...] would clearly be in conflict with human activity, raising questions about the feasibility and diverse social implications of this strategy."[83] The magnitude of possible dispossession and displacement implied by these conservation goals leads the authors to state that while they "recognize the importance of conserved areas for the future of life on Earth," they urge advocates of Half-Earth, 30x30, Global Deal for Nature, and similar ambitious area-based conservation targets to "consider important broader issues, such as environmental justice, the multiple values people attribute to nature and the need for action to tackle the ultimate economic consumption and production drivers of biodiversity loss."[84]

Given the shocking magnitude of these proposed policies, it should not be particularly surprising that Indigenous groups and their allies have denounced the 30x30 slogan as neocolonial. As the World Rainforest Movement (WRM) put it in a May 2020 bulletin, while big conservation organizations carefully promise to align conservation activities with human rights and with community- and participatory-based approaches, "none of the proposals that aim to make conservation appear more people-friendly have gotten to the core issues of who controls land in Protected Areas, or who decides whether a location is decided to be protected, and what that means."[85] WRM's basic question opens out onto many others: Are decisions about which places are declared PAs made by scientists working with BINGOs and corporations? Are all of these people and organizations based in the Global North? What role do governments in Global South nations have? To what extent are non-Western forms of collective customary land rights recognized when such decisions are made? As these questions suggest, there is a crisis in models of governance for the planetary environmental commons. Dominant conservation organizations have an answer to this crisis: Nature-based Solutions. Yet, as we shall see, these so-called solutions are fatally flawed and failing.

## "Nature-based Solutions" and Accumulation by Extinction

In a large lecture hall at the University of KwaZulu-Natal, I joined a group of activists gathered for a role-playing game. Along with thousands of other climate justice activists from around Africa and the rest of the world, I had traveled to Durban, South Africa, in 2011 to attend a counter-summit to COP17, the United Nations Climate Change Conference. We spent a week swapping stories and information about one another's struggles and about the ways in which corporate polluters and powerful states were finding ways to duck their responsibility to cut carbon emissions. Now we were gathered in a lecture hall to play a game designed to help us understand one such approach, a strategy known as "carbon offsetting."[86] Carbon-offsetting schemes such as the UN's program on Reducing Emissions from Deforestation and Forest Degradation (REDD) had dominated the previous year's UN Climate Meeting in Cancún, Mexico, where World Bank president Robert Zoellick spoke enthusiastically in favor of REDD.

As we sat in the lecture hall, an organizer walked around the room holding a hat, inside of which were many slips of folded paper. I picked one and unfolded it, to find that I had been assigned the role of a resident of Richmond, California, a community of working-class people of color located adjacent to a Chevron oil refinery. Others in the group had drawn slips with roles such as Chevron executive, corporate lawyer, representative of the California EPA, Indigenous resident of the Lacandon rainforest in Chiapas, Mexico, governor of Chiapas state, and rubber tapper in the Amazon rainforest. The organizers quickly explained the official view about how carbon offsetting works: the world needs to find a way to mitigate the carbon being emitted into the atmosphere; we also want to maintain the biodiversity found in the forests of Global South nations; REDD and other carbon-offsetting schemes allow polluting industries to pay poorer countries to maintain or even expand their carbon-absorbing forests; this gives polluting industries in the Global North time to find technical ways to slash their emissions. We were then each given five minutes to write a short statement from the perspective of the person whom we'd picked about how this carbon-offsetting system would affect them. The organizers gave each of us two minutes to deliver this speech. The results were simultaneously entrancing and infuriating.

Activists playing the roles of corporate executives, lawyers, and government officials gave speeches that echoed the duplicitous rhetoric heard every day inside the carefully policed doors at the nearby UN climate negotiations. They talked about win-win solutions, and about the cost-effective nature of carbon offsets, which help developing nations by providing much-needed investment money. One activist, perfectly capturing the rhetoric of particularly oily officials, even talked about the new green economy and about the wonders of the market, which would lead to the speedy commercialization of genetically engineered trees to absorb carbon and produce biomass energy.

These speeches were interrupted by a wave of upsetting information delivered from subaltern environmental perspectives. Those of us playing the roles of ordinary people in both Global South and North nations talked about how carbon offsets let polluting industries off the hook, allowing them to keep polluting the land and air, in the process destroying people's lives. I knew about the high asthma rates suffered by communities subjected to environmentally unjust forms of pollution, and did my best to denounce the Chevron refinery's desire to keep pumping toxins into the air. I learned a great deal from people playing the roles of Indigenous activists in Mexico and Brazil, who talked about how they were being pushed off their ancestral land by government officials, who in turn gave concessions to big European or US lumber companies that planted fast-growing eucalyptus trees designed to be cut down only a few years later for industrial uses. Whatever paltry amount of carbon these trees absorbed during their short lives was circulated back into the atmosphere as soon as they were harvested. One particularly angry person playing a Mexican farmer said that carbon offsets are like trying to lose weight by paying someone else to diet for you. The whole idea of carbon offsets is so crazy and yet so diabolical that it sets you crying and laughing at the same time.

It can be no coincidence that the idea of Nature-based Solutions (NbS) appeared at exactly the same time as the offsetting schemes that climate justice activists decried at this COP counter-summit. In fact, as Fiore Longo of the pro-Indigenous rights group Survival International explains, the term *Nature-based Solutions* made its debut in a 2009 document the IUCN wrote for the UN climate negotiations conference.[87] This timing makes it clear that global elites were becoming aware not just of the increasingly dire material impact of climate chaos, but also of increasingly strong movements for climate justice around the world. As I discussed in an article I wrote in 2010, the movement for climate justice had articulated strong demands over the previous decade and was engaging in increasingly militant mobilizations, a trend that culminated in the 2010 World People's Conference in Bolivia.[88] Elites clearly felt they needed to respond somehow to the intertwined threats of climate and political crisis. Their answer: fake green capitalist fixes like carbon offsetting and other NbS. Yet, as the speakers at the counter-summit in Durban made clear, these green capitalist solutions will never solve the crisis generated by an economic system that must expand ceaselessly or collapse. In fact, it turns out that NbS actually intensify the intersecting social and environmental contradictions that characterize the climate crisis. And while NbS might superficially seem to differ from colonial conservation, these approaches end up fostering the same kinds of exclusion that characterizes protected-area policies.

According to the IUCN, "Nature-based Solutions are actions to protect, sustainably manage, and restore natural and modified ecosystems in ways that address societal challenges effectively and adaptively, to provide both human well-being and biodiversity benefits. They are underpinned by benefits that flow from healthy ecosystems and target major challenges like climate change, disaster risk reduction, food and water security, health and are critical to economic development."[89] This definition clearly reflects a shift in the core doctrines of capitalism, with nature no longer being viewed simply as what economists call "an externality," something to be ignored or, at best, used as a dump for the toxic by-products of industry. Economic systems are now seen as nestled within a natural world that must be sustainably stewarded. If this is done correctly, according to the IUCN, nature can provide "benefits" to both humans and biodiversity, and, conveniently, it can also address some of the environmental crises produced by a system hell-bent on frenetic growth.

But notice that the statement concludes with an assertion that all of this helps to further "economic development"—or, more bluntly put, the continued functioning of the capitalist system. Like the UN's Sustainable Development Goals, in other words, NbS are defined by a core contradiction: they are an effort to cease degrading the environment and to achieve harmony with nature, and, simultaneously and impossibly, a call to continue global growth rates of 3 percent in order to achieve human development objectives. Yet, as researchers such as Jason Hickel have shown, such a global growth rate will render it empirically infeasible to achieve reductions in aggregate global resource use and to reduce $CO_2$ emissions rapidly enough to stay within the carbon budget to limit warming to 2°C.[90]

Despite its carefully crafted image as the key means for saving the planet, green capitalism is engaged in what could be termed *accumulation by extinction*. As the geographer David Harvey has explained, the capitalist system tries to manage its crises through "spatio-temporal fixes."[91] Capitalists have to find a place to put the treasure they've hoarded through successful competition. Failure to do so would mean their capital might be seriously devalued, a wrenching experience that economists term a *recession* or, worse still, a *depression*. Capitalists evade such dangerous economic and political crises by investing their money in new places, using what Harvey terms processes of "spatial reorganization" such as urban "development," which perpetuates mass displacement of poor urban residents, as well as through violent forms of geographical expansion such as imperialism. These kinds of "fixes" also kick the can of crisis further down the road, since the new sector where capital is invested eventually grows into a bubble of its own that threatens to burst and drag down the larger economy. The extreme inflation of the housing market around the globe today is a prime example of one of these spatio-temporal fixes: global elites have plowed the mammoth assets they have stripped from ordinary people during the neoliberal era back into luxury condos in the sky, making housing unaffordable for ordinary people and threatening to provoke a crash, as the worth of all these over-valued assets must be slashed dramatically.

Despite being faced with looming environmental breakdown, contemporary capitalists are doing everything they can to continue accumulating wealth. Indeed, capital accumulation is speeding up and intensifying in the teeth of the environmental crisis. According to the United Nations, for instance, the global material footprint rose from 73.2 billion metric tons in 2010 to 85.8 billion in 2017, an increase of over 17 percent.[92] While ramping up production and consumption, capitalists also seek to invest in sites around the world that are likely to absorb not just over-accumulated capital but also waste products of production such as greenhouse gases, plastics, clothes, and other residues of the capitalist system. This is what the conservation industry is increasingly specializing in: it sets aside PAs that are supposed to conserve nature, and, conveniently, to absorb carbon emissions, allowing the capitalist system to carry on at full tilt.

Yet rather than leading to meaningful mitigation of emissions in particular and of the environmental crisis in general, such efforts to shift the crisis around in space and time are decimating the biosphere. Notwithstanding decades of global conservation work, annual high-level climate conferences, and proliferating rhetoric about greening capitalism, carbon continues to accumulate in the atmosphere and species still go extinct at ever-increasing rates. If conservation is a kind of spatio-temporal fix for the built-in contradictions that plague capitalism, it is one that actively produces extinction.

As is true for carbon-offsetting programs, an embrace of NbS allows polluting corporations like Shell Oil—a big supporter of NbS—to keep emitting greenhouse gases, but at the same time to claim it is cutting its aggregate emissions by supporting the creation of PAs that stock an amount of carbon dioxide equivalent to what it is emitting.[93] In many instances, big corporations can claim that they are actually helping to clear the atmosphere of polluting gases (while they continue on with business as usual) by funding the establishment of tree plantations. This is what "net zero" really means: in the words of that activist from Durban, you claim you are not gaining weight because you are paying someone else to go on a crash diet, even while you continue to gorge yourself on the fat of the land.

The creation of exclusionary PAs—i.e., colonial conservation—is absolutely central to contemporary green capitalist efforts to address the climate crisis. NbS was, for example, a key topic at COP26 in Glasgow in 2021.[94] Citing 2017 estimates, Fiore Longo explains that afforestation projects account for nearly half of the estimated potential for climate mitigation through NbS.[95] But absorbing such a vast quantity of carbon would require planting trees over an estimated 700 million hectares, an area of land almost the size of Australia. Where, Longo asks, would this land be found? Certainly not, she quips, in France or the United Kingdom, key supporter countries of NbS. The upshot, Longo states, is that Indigenous peoples and other forest-dependent communities—the people least responsible for the climate crisis—face the threat of a new round of evictions from their land. And these policies of violent enclosure and eviction are not even likely to work on their own terms: Longo cites a 2019 study that found, after examining more than twelve thousand PAs across 152 countries, that such conservation reserves have done virtually nothing over the last fifteen years to reduce human pressure on wildlife.[96] The PAs themselves have also been subject to the negative environmental and social impacts associated with mass tourism, as well as trophy hunting, logging, and mining.

The level of corruption, cynicism, and brutality involved in contemporary NbS efforts is shocking. Take TotalEnergies's NbS project in the Republic of Congo. As Simon Counsell explains in his account, TotalEnergies—the world's seventh-largest producer of oil—announced a project in 2021 that would plant a "40,000 hectare forest" in Congo, a forest which would "sequester more than 10 million tons of $CO_2$ over 20 years."[97] Two years prior to this, TotalEnergies had announced the creation of an NbS unit to develop "natural carbon sinks" to "sequester $CO_2$ from its operations."[98] But that same year, TotalEnergies had acquired an oil exploration permit for a 1.5 million-hectare plot in Congo, in what was revealed to be one of the world's largest peat deposits. Peat is an extremely carbon-dense ecosystem, so drilling for oil there would constitute an environmental damage redoubled. At the same time, the Central African Forest Initiative (CAFI) announced that $65 million was being made available to Congo for "the protection and sustainable management of peatlands in the Republic of Congo."[99]

But the plot thickens further, for it turns out that portions of the environmentally sensitive Bateke Plateau where the TotalEnergies oil concession is located are managed by the New York–based Wildlife Conservation Society. Instead of denouncing TotalEnergies, the Congolese government, and the corrupt CAFI grant, the WCS proudly claims that it is working with the government of Congo to map the plateau's biodiversity, with the goal of eventually creating a PA there, to be called the Bambama-Lekana Park.[100] The upshot for local people is never mentioned, but it's worth remembering that an inquiry focused on the northeastern edge of the Bakete Plateau by the Rainforest Foundation UK found that—as elsewhere in the Congo Basin—customary community lands are essentially contiguous. As Counsell argues, this suggests that the entire area of the TotalEnergies concession is claimed under customary tenure, and is used for collecting, hunting, and subsistence farming. Since the WCS has already been involved in evictions from other national parks it has established in the Congo, there's no reason to assume that this new park would not simply continue the long tradition of establishing exclusionary PAs.

All of this might come as a total shock if one were not aware of the long colonial genealogy of conservation. The idea of NbS is just another turn of this colonial conservationist screw. Indeed, as we have already seen, the concept of NbS was originally developed by the conservation industry—including the WCS. As Counsell argues, the scheme was concocted as a means to sell carbon credits from PAs in order to generate funds to establish more PAs. Never mind that schemes such as the Bambama-Lekana Park and the broader nature-based bait-and-switch scheme of which it is a part are unlikely to have any positive impact on climate change, and are almost certain to be a social and environmental crime of epic magnitude. After all, any positive climate impact resulting from the park will be more than counterbalanced by the carbon emissions resulting from oil extraction nearby. NbS such as TotalEnergies's Congo dodge are carefully crafted fictions, calculated to induce what might be called, drawing on cultural critic Lauren Berlant, "cruel environmental optimism": a frame of mind that is intended to rout climate justice activism by making people believe that capitalism is capable of addressing its grievous damage to the world, even while toxic corporations keep killing the planet.[101]

## Decolonize Conservation!

Indigenous peoples and other forest-dependent communities have always resisted fortress conservation. In their ecological history of India, for instance, Madhav Gadgil and Ramachandra Guha describe the reactions of different groups in India to the dispossessing effects of colonial forest and game laws. When people's long-standing relationships with forests were made illegal by these laws, they at times lost their autonomy and were forced into relations of quasi-serfdom to the British forest administration or to more powerful cultivating castes.[102] In some instances, small tribes whose hunting activities were made illegal literally ran away, successfully migrating to other areas. Some even turned to banditry. In the cases of larger tribes whose practices of shifting cultivation in forests had been rendered illegal, they took to violent confrontation with British authorities, including the invasion of reserved lands and their clearing in preparation for cultivation.[103] When men of the Saora tribe in the state of Odisha were jailed for such behavior, women of the tribe continued to cultivate the cleared land. In some cases, such resistance even grew into open conflict with the colonial state. In perhaps the most dramatic instance of such overt resistance, tribal peoples from the hilly forested areas in the southeastern state of Andhra Pradesh rose up in what came to be known as the Rampa rebellion of 1879–80. Arguing that "as they could not live they might as well kill the constables and die," rebels attacked and burned several police stations and even executed a constable.[104] Several hundred policemen and ten army divisions were needed to quell the uprising.

As we have seen, there is a direct link between the modes of exclusionary conservation practiced in the colonial period and fortress conservation today. It should therefore be no surprise that resistance to the enclosure of collectively owned land continues around the world. There have been significant victories in this struggle. In 2006, for instance, the Indian government passed the Forest Rights Act. At a stroke, this measure turned much of the nation's forests into a community rather than a state asset. In doing so, it reversed the long history of colonial and postcolonial enclosure and dispossession of forests and other communally owned lands in India. If fully implemented, the law could enshrine community rather than state control of over half the country's forests.[105] The magnitude of this change is truly epic: over half of India's rural households depend on community or common lands for their livelihoods.[106] The Forest Rights Act held the potential to be the largest land reform in India's history, and one of the largest in the world. Its passage was a massive victory in the struggle of Indigenous peoples and other land- and forest-dependent communities for their rights to the commons. Winning adequate implementation of the law has not, however, been easy. Since the act's passage, the Indian government has announced plans to privatize large portions of its forest estate, and many Indian state administrations are dragging their feet in carrying out the handover of forests they are tasked with organizing.[107]

The mix of victories and setbacks that characterizes struggles for forest rights in India is similarly evident around the world. In other nations such as Indonesia and Brazil, for instance, Indigenous groups have won significant legal victories against the enclosure of common lands but have yet to see these victories bear fruit on the ground. In Indonesia, for example, an action by the Indigenous People's Alliance of the Archipelago (Aliansi Masyarakat Adat Nusantara, or AMAN) in 2013 led the country's Constitutional Court to annul government ownership of the country's forests. This ownership had previously been won through the US-backed Suharto regime's bloody counterinsurgency operations over decades. The court ruled that "members of customary societies have the right to . . . use the land to fulfil their personal and family needs."[108] Nevertheless, in 2015 the Indonesian government announced plans to fast-track extensive land procurement to support its sweeping infrastructure goals. And elsewhere in the world, the economic crisis of 2008 sparked a fresh round of land grabs, challenging and in some cases reversing decades of struggle by Indigenous people and local communities to win their land back.[109]

From a global viewpoint, the injustices faced by Indigenous peoples and local communities fighting for the right to land are stark. According to the 2016 Oxfam report *Common Ground: Securing Land Rights and Safeguarding the Earth*, the claims of customary land users cover more than 50 percent of the world's land area, yet legal recognition of ownership is restricted to just 10 percent of lands.[110] Worse yet, such official recognition is limited to just a handful of countries: China, Brazil, Australia, Mexico, and Canada. And even in these countries, the real situation is far less progressive and just than it appears on paper. In Australia, for example, although some areas are formally subject to Aboriginal ownership or control, common law rights have been effectively extinguished by the legal arrangement known as "native title," which requires Aboriginal communities to show a continuous connection to the land—despite generations of forcible removal of people from their lands.[111] We have seen how such policies have developed over centuries in a country like India. This situation often leaves Indigenous communities unable to resist extractive projects launched on their traditional lands.

The gap between customary rights and legal title is greatest in sub-Saharan Africa. Here, as in India, rural villages consider the forests, pastures, and other natural resources that surround them to be their collective property. Yet only 3 percent of land is legally recognized as owned by Indigenous peoples or local communities.[112] Land actually ceded by governments is often located in extreme environments such as deserts or mountainous areas. This systemic dispossession helps explain the rampant injustice that TotalEnergies has been able to perpetrate in Congo. When Indigenous peoples and local communities do win control of land, they are able to exclude outsiders from claiming community lands. In cases where the land is expropriated, they can obtain due process and compensation when government authorities make concessions to extractive development.

It is essential that the rest of the world comes to recognize the environmental stewardship of the world's 2.5 billion Indigenous peoples and local communities. These groups must be granted the land rights they are demanding if we are to find genuine and just ways out of the conjoined biodiversity and climate crises. This is perhaps the foundational argument of the *People's Manifesto for the Future of Conservation*, the document published at the Our Land, Our Nature! counter-summit to the 2021 IUCN conference in Marseille. The *People's Manifesto* argues that "the only sustainable, just, and real solutions" to the biodiversity and climate crises "lie with humanity—in particular, with Indigenous Peoples and local communities, who are the best guardians of biodiversity, and with a model of conservation that puts human rights and human diversity at its center."[113] Dominant models of conservation—those fostered by the conservation industry—only give lip service to community inclusion, the *People's Manifesto* argues, and remain dominated by a model that is far from being rights-based. According to the manifesto, reforms such as people-centered conservation are "only cosmetic" and only include Indigenous and local people as an afterthought.[114]

For the groups and organizations that gathered to author the *People's Manifesto*, genuine conservation would involve not just recognizing the environmental stewardship of Indigenous people and local communities. Decolonial conservation would also fight against the real causes of environmental destruction: the overconsumption and exploitation of resources led by the Global North and its corporations. The *People's Manifesto* here echoes the Cochabamba Declaration by arguing that preventing planetary ecocide means challenging and reining in contemporary forms of green colonialism carried out by the world's core imperial nations. To do so, social movements in both the Global South and North must insist on the necessity of dismantling the unnecessary, polluting sectors of production and consumption that are currently integral to the economic and social model of wealthy capitalist nations. As the philosopher Kate Soper has argued, the cult of endless growth and the deluge of cheap, disposable commodities that it generates is not only destroying the world but also profoundly damaging those who reside in these wealth-afflicted societies.[115] Capitalism is a world-system: it may first affect certain zones demarcated for sacrifice, but it will ultimately annihilate the entire planetary commons. As the activist Ashish Kothari put it at the Our Land, Our Nature! counter-summit:

> Suppose you magically manage to protect 30 percent of the Earth: you kick out everybody and it's all protected for wildlife. Sounds good. If the rest of the 70 percent or 50 percent is going to continue with business as usual, how will you protect what's within the 30 percent? Climate change does not know boundaries. Pollution does not know boundaries. Toxic products don't know boundaries. They are going to impact wildlife even if you have a strictly protected 30 percent portion of the Earth. So unless fundamental alterations are

made in the economic growth-based neoliberal capitalist models of development, we have no hope whatsoever of protecting biodiversity on the planet.[116]

Activists must resist efforts to "solve" the biodiversity and climate crises by shifting the burden onto the backs of the world's poor, through NbS that allow powerful corporations and their government stooges to carry on with business as usual.

The *People's Manifesto* includes a series of demands foundational to decolonial conservation. Most basically, this includes the demand that governments "fully respect Indigenous Peoples' land and forest rights, including their collective customary land and forest use."[117] This means that there must be a total cessation in the construction of all protected areas that exclude Indigenous and local communities, and that no such PAs be constructed in the future without the full free, prior, and informed consent of the people on whose customary lands the PAs will be situated. This demand flows directly from the rights won and guaranteed in the UN Declaration on the Rights of Indigenous Peoples.

Linked to these indictments of contemporary dominant models of conservation is an insistence that governments and big conservation organizations acknowledge the toll that fortress conservation has taken historically, and, what is equally important, "make concrete plans for reparations, including land back." As Olúfẹ́mi O. Táíwò has argued, such demands for land back must be seen as part of a broader movement for reparations that ties together labor, climate justice, migration, and land restitution as part of a worldmaking project designed to reverse the unsustainable global structures built of colonialism and racial injustice.[118]

The *People's Manifesto* thus connects the call for decolonial conservation to a global demand for the restitution of land rights. This restitution of land rights to Indigenous peoples and local communities must take place in virtually all the nations of the world. Movements can strengthen their work in particular national contexts by placing their demands within the context of global solidarity. Thus, for instance, a recent call to return national parks to Native American tribes in the US that was perceived as quite radical seems less so when compared to developments such as the

Forest Rights Act in India or the sweeping restitution that courts have approved in Indonesia. As David Treuer argues, "All 85 million acres of national-park sites should be turned over to a consortium of federally recognized tribes in the United States. . . . To be entrusted with the stewardship of America's most precious landscapes would be a deeply meaningful form of restitution."[119] There is a growing global movement for land back, and conservation efforts need to be part of that movement.[120]

The *People's Manifesto* closes with a set of demands that lay the groundwork for a decolonized conservation that is simultaneously anti-capitalist and anti-imperialist. The manifesto does this by calling on rich countries to provide the financial resources necessary to fund land back efforts in the Global South. The rich imperial nations must also agree to genuine reductions (not just shifts) in extraction, and enforce this on corporations based in the imperial core. These demands are key given the context of a fresh round of land grabs over the last decade or more, as well as the danger of new forms of extractivism triggered by the voracious need of Global North nations to secure new raw materials for the energy transition. After all, as Asad Rehman has warned, the Green New Deal championed by political progressives in wealthy nations like the US and UK runs the risk of becoming another imperial resource grab unless strong curbs are put on extraction from poorer countries, preferably by those countries themselves.[121] Finally, the *People's Manifesto* demands that rich nations abandon fossil capitalism and stop pretending that NbS will solve the biodiversity and climate crises while they continue depleting the global environmental commons as if there were no tomorrow.

Billions of people around the world are already responsible for conserving biodiversity, building on the efforts of their ancestors to live harmoniously with the environments they inhabited. All too often, these Indigenous people and local communities have been and still are marginalized and persecuted by dominant forms of conservation. And yet they endure, they steadfastly protect their sublime natural inheritance, and they continue to fight against new forms of colonialism. As Pranab Doley, the activist from the region of India where the Kaziranga National Park is situated, put it during his testimony at the Our Land, Our Nature! counter-summit, "The fight is against this capital-intensive idea, where they're trying to commodify everything, not only nature but sheer human existence, and to use it for profit. We stand strongly against this. I am honored to share experiences from the place I come from. I and a lot of other people are carrying on a daily existence—it's a constant fight of millions of people across the world, and I want to embody it."[122]

# CLIMATE DEBT AND BORDER ABOLITION

C ruz Martinez began a hunger strike in March 2023. Two weeks into the strike, wracked by hunger pangs, Martinez, then twenty-two, said that he intended to carry on his protest "until I drop."[1] Martinez was among forty-five people coordinating a hunger strike that month against the harrowing conditions in two immigrant detention centers in California run by the GEO Group, a private prison company contracted by US Immigration and Customs Enforcement (ICE). Martinez explained that he and his fellow detainees in these facilities were fed rotten food, were forced to work for "slavery wages" of one dollar a day, and faced long waits for medical treatment. Worst of all, though, was the uncertainty about their fate resulting from the lack of due process while in custody. "I've never been so hungry in my life," Martinez said, "but we want to be with our families."[2]

Almost simultaneously with the protests in California, three hundred detainees at the Central Louisiana ICE Processing Center, also run by the GEO Group, began a hunger strike against conditions in that facility. Like Martinez and his comrades in California, the protesters in Louisiana were demanding not just an end to the rampant neglect and unsanitary conditions in the facility but immediate release from detention as well as transparency from ICE regarding individual cases, including case status, future court dates, and release requests.[3]

The outbreak of independent hunger strikes at different ICE detention facilities across the United States is no mere coincidence. In a sprawling system where approximately 250,000 people are held in custody by ICE, flagrant abuse of detainees is widespread. The entire system of immigrant detention is unjust and inhumane, based on incarcerating immigrants while they wait for a determination of their immigration status or potential deportation. People held in ICE jails can be documented or undocumented immigrants, they can be survivors of torture, people seeking asylum, visa holders, people who have been granted the permanent right to live in the US, people who have lived here for years and may have spouses and children who are citizens, individuals with mental health or medical conditions and other vulnerable groups, including pregnant women, families with children, and even babies.[4] Not only are people incarcerated in ICE detention centers denied their liberty, they are also denied access to lawyers, separated from their families and loved ones, and subjected to severe medical neglect. Detainees are held in a sprawling network of two hundred jails across the country, many of which are run on a contract basis by local and state authorities and by for-profit prison companies like CoreCivic and the GEO Group. This contracting arrangement creates a perverse financial incentive to keep people locked up, since private prison companies and local governments make money off detainees through lucrative contracts with ICE.

Extreme human rights abuses take place frequently in this network of ICE prisons. Verbal, physical, and sexual assaults against people held in ICE detention are commonplace.[5] In addition, isolation is still common in US prisons and detention centers, even though the United Nations has called prolonged solitary confinement a form of torture and has recommended that it be banned.[6] Such abusive practices continue today, despite Joe Biden's pledge to end these forms of inhumane treatment while he was on the campaign trail in 2020. In a 2021 report, the Center for Victims of Torture—the oldest and largest torture survivor network in the US—found that ICE "systematically exposed detained migrants to violations of the prohibition on torture and other cruel, inhuman, or degrading treatment or punishment."[7]

ICE's brutal incarceration policies don't just confine people in one site. The agency also forcibly moves people around from one jail to another, a policy ICE calls "circular transfers." Ernest Francois is one of many detainees who have been subjected to this cruel practice. A forty-nine-year-old man from Haiti, Francois was detained by ICE at Essex

County Correctional Facility in Newark, New Jersey, in 2017. In 2020, in need of surgery for a back injury, Francois joined with a group of a dozen other detainees to speak out against medical negligence in the facility. He immediately began suffering retaliation for challenging authorities in the jail, abuse that included solitary confinement, physical and sexual abuse, and racist and homophobic threats.[8] In April 2021, Francois submitted a complaint to Essex County Commissioners in New Jersey regarding the treatment he received while detained there. After his case received extensive media coverage, ICE authorities transferred him to a nearby prison. There, Francois again began organizing with other detainees against the conditions, and ultimately participated in a mass hunger strike in September 2021. Several months after his first transfer, Francois was transferred again, this time to a detention center in Florida.

Transfers like those to which Francois was subjected seem intended to move people as far as possible from their family and attorneys. In addition, according to reports from detainees collected by the organization Freedom for Immigrants, transfers are a form of retaliation by ICE against people who, like Francois, speak up against the conditions of their detention.[9] Transfers rupture the collective solidarity detainees have built. What's more, the process of moving from one detention center to another often involves grueling physical conditions, including physical violence, extended social isolation, and a lack of access to essential resources like food, water, and medications. They amount, in other words, to a form of torture.

People subjected to repeated transfers also reported a feeling of lost autonomy and a sense of being "trafficked" by ICE, moved from one site only to be put to work immediately upon arriving at another one. As Francois put it, "It's like we ain't nothing but slaves. . . . They move us around without telling us anything, and they make us work for free. It's mind-blowing that they think they can do this to a human being."[10]

The growing abolitionist movement has achieved remarkable success in challenging the kinds of abuses that migrants like Francois have endured in recent years. As *Building Power*, a report by Elena Hodges of the Immigrant Rights Clinic at NYU Law School, outlines, abolitionist organizers and advocates have helped terminate thirty-six ICE detention facility contracts since 2017.[11] Transfers remain an effective weapon against abolitionist organizers and advocates. Instead of releasing detainees when the movement secures a victory and shuts down a detention center, ICE regularly transfers detainees elsewhere. Some policymakers and media organizations then blame the abolition movement for the resulting social dislocation of immigrants. Yet these efforts to discredit abolitionist victories underline the movement's power.[12]

Ernest Francois's history attests to the gathering strength of this abolitionist movement. The public exposure surrounding his complaints about his treatment helped push Essex County to terminate its detention contract with ICE, even as Francois was transferred to another facility. That summer, he was named as a complainant in a case brought by the Center for Constitutional Rights about abusive practices in the Bergen County Jail. Within a few months, the ICE detention contract at this jail was also terminated. Francois continued to lead collective protests inside detention facilities. After being told by a sheriff in a detention facility in Glades, Florida, that a rope would be placed in his cell to be used as a noose, Francois worked with allied community members, the media, and politicians to win his release on bond in October 2021. He was able to return to his family in New York with help from the abolitionist organizations that had fought from outside prison walls.

⤙

Ernest Francois's and Cruz Martinez's stories offer a window onto the oppressions to which contemporary migrants are subjected—and onto the courageous resistance of migrants that is at the heart of the abolition movement. The right to move is also central to movements for environmentalism from below. Indeed, border abolition and the defense of refugees and migrants must be regarded as a key element of the climate justice movement. This means, first and foremost, challenging the racist legacy of the environmental movements in core imperial nations. From the eugenicist thought of early environmentalists like Teddy Roosevelt to the kinds of enduring colonial conservation practices that I discussed in the last chapter, the dominant environmental movement has long been shaped by patriarchal, white supremacist logics. Environmentalism from below works to undermine and overthrow these oppressive traditions. As we will see, this task is made all the more urgent by the reanimation of toxic legacies by contemporary green nationalist and eco-fascist movements. Building solidarity between the border abolition and climate justice movements is key to resisting such forms of racist environmentalism.

Fighting these dangerous movements also entails recognizing the rights of climate change–displaced peoples who are currently legally invisible. The United Nations Refugee Convention of 1951 recognizes people fleeing political persecution as eligible for refugee status and asylum but says nothing about the rights of those displaced by environmental disasters or the slower-onset impacts of climate change.[13] The system of international law holds that it is unable to identify a perpetrator of the crimes from which climate refugees are fleeing, so it refuses to recognize the status of climate refugees. This must change: climate refugees must be recognized by national and international legal bodies, and must be given support by those who are responsible for the climate crisis.[14]

Migrants and refugees are certainly visible currently, but not in ways that would help garner them rights—let alone in a manner that would abolish the incarceration and deportation regime in core imperial countries like the US and EU member states. Activists like Cruz Martinez and Ernest Francois are rebelling against the oppressive and exploitive conditions in ICE detention facilities within a political context in which public discussion of a migration "crisis" has become virtually ubiquitous. Like the figure of the "mugger" in the 1970s, who for Stuart Hall and the other authors of *Policing the Crisis* became a symbol of the breakdown of the consensus that had supported the post-1945 Keynesian welfare state, the migrant and refugee today are invoked to articulate a broad set of crises that extend far beyond migration per se. The components of the "crisis" for which the migrant stands in include the offshoring of working-class jobs, galloping economic inequality produced by the financialization of capitalism over the last half century, and the increasing cultural cosmopolitanism introduced by the very communication technologies that have helped facilitate a new international division of labor.[15] Immigration has become a "funnel" issue for the far right, since all other issues can and are subsumed within the call to halt and ultimately reverse the arrival of nonwhite foreigners.[16]

Never mind that the threat of deportation that hangs constantly over the heads of migrants and refugees in wealthy nations like the US and the European Union is a highly convenient tool of the ruling class in these countries. The ever-present threat of expulsion works to discipline immigrant labor, which in turn depresses wages.[17] But while migrant and refugee inclusion within labor markets plays an important function in contemporary capitalism, it is also true that many refugees are not put to work but are kept stalled indefinitely in camps and other transit points. Nonetheless, even when trapped in such states of indefinite suspension, refugees have an economic function for the capitalist system. They are, for instance, often included in circuits of credit and debt through the issuing of electronic vouchers for services and humanitarian credit cards. In addition, the surveillance and tracking of refugees is big business for tech companies and the security state, which seek to thereby predict future migration routes.[18] The result is policies of immigration interdiction that grow increasingly deadly.

But the threat of detention and deportation goes beyond a narrow economic rationale. The detention and deportation regime is a powerful weapon in the arsenal of the imperial nation-state, which uses these strategies as a political instrument to contain dissent and suppress political movements.[19] The figure of the migrant/refugee as menacing alien presence also conveniently helps produce an image of a civilized "homeland" surrounded by various failing and anarchic states. In core capitalist nations like the US and member states of the European Union, public discourse has become increasingly suffused by the kind of victim mentality that characterized white settlers in South Africa, who protected themselves during their trek into the African interior by forming their wagons into a fortified circle, or *laager*. In the hands of race-baiting public figures like Donald Trump and his many epigones and followers on the far right, the figure of the migrant and the refugee—who are effectively collapsed into one—generates a myth of the ethnically pure nation besieged by people fleeing infernal zones of social breakdown. This conveniently elides the role of the US and Europe in the violent invasions, clandestine wars, debt-produced instability, and other colonial and postcolonial atrocities that have destabilized the areas from which most migrants/refugees flee.

While xenophobic rhetoric has been a staple of right-wing politics ever since the economic and political crises of liberal capitalist states in the 1970s, in recent years it has become increasingly linked to an acknowledgment of environmental crisis. In this nascent brand of eco-fascism, right-wing ideologues are turning away from the climate change denialism of an older generation of reactionaries to argue that the climate crisis is menacing scarce environmental resources. The resulting "green nationalism" is predicated on a Malthusian version of the *laager* doctrine, in which white supremacists argue that immigrants and people of color more broadly should be banned from entry and deported, so that natural resources remain available to the select (white) few.[20]

Among the many disturbing examples of this demographic conspiracy theory is the rambling manifesto of Brenton Tarrant, the perpetrator of the 2019 killing of fifty-one people in Christchurch, New Zealand. Tarrant's screed draws on populist rants such as the Frenchman Renaud Camus's essay "The Great Replacement," which argues that European elites are seeking to replace white Europeans through mass migration. But Tarrant gives these ideas a chilling eco-fascist spin, arguing:

> The environment is being destroyed by over population, we Europeans are one of the groups that are not over populating the world. The invaders are the ones over populating the world. Kill the invaders, kill the overpopulation and by doing so save the environment.[21]

Tarrant's manifesto builds on the kind of eugenicist thinking that undergirds Garrett Hardin's infamous 1968 essay "The Tragedy of the Commons" with its ahistorical and incorrect analysis of commons governance traditions and its eugenicist attacks on the welfare state for encouraging the poor to reproduce.[22] Tarrant's manifesto, symptomatic of an all-too-widespread genre, lays bare how fascists are seeking to take advantage of the intensifying climate crisis by reviving the vile racist fear-mongering of eugenicists like Hardin.

To call Tarrant's manifesto a horrific racist distortion would be an understatement. The idea that these migrants and refugees are menacing the core imperial nations is ludicrous. It is certainly true that the number of refugees and migrants are at an all-time high: in 2022, the number of forcibly displaced people surpassed one hundred million for the first time in world history.[23] And climate-induced migration is an increasingly pressing reality. The speedy and severe onset of the climate crisis has led to a dramatic increase in the number of climate-related disasters over the last twenty years. There were, as a result, 32.6 million people displaced from their homes by natural disasters in 2022.[24]

But these people were "internally displaced," in the parlance of the international legal system, meaning that they remained within the countries where they experienced natural disasters severe enough to force them to flee their homes and communities. While they suffered many of the same dangerous circumstances as refugees, including being forced from their homes and communities, they did not cross an international border into any of the rich nations. Indeed, of the stark figure of 100 million global refugees and migrants, nearly two-thirds (71.1 million) are internally displaced people. Many of them have ended up struggling to eke out an existence in the densely packed cities of the Global South that I discussed in chapter 2.[25]

In addition, low- and middle-income countries host a full 74 percent of the world's refugees and other people who qualify for international protection.[26] Furthermore, of the world's 32.5 million refugees, 69 percent are harbored in countries neighboring their countries of origin. Only one wealthy country—Germany—is even in the list of the top five countries hosting the largest number of refugees.

The weight of the climate crisis is, in other words, falling most heavily and overwhelmingly on those who are least well-resourced to cope with it. These people and places are also least responsible for the climate crisis. As a recent report by the organization CARE explains, "Despite the fact that the poorest 50% of the world's population is responsible for just 7% of global emissions, developing countries will face 75–80% of the costs of climate change."[27]

The conditions that produce these forms of displacement are set to get much worse. According to the seminal 2020 paper "The Future of the Human Climate Niche," by 2070 between one and three billion people are projected to live outside the climate conditions that have sustained human life for six thousand years.[28] The authors' analysis shows that in a "business as usual" climate scenario, the mean annual temperature will rise to an unbearable 29°C (84°F) in the one-third of the planet that is currently inhabited. Only about twenty million people currently live in regions where the average temperature exceeds 29°C, an area mainly near the Sahara Desert that presently constitutes less than 1 percent of the earth's land. Yet as the world gets hotter because of global warming in the coming decades, huge swaths of Africa, Asia, South America, and Australia will be in this same temperature range. With warming of 3°C, conditions akin to those in the Sahara today are predicted to envelop 1.2 billion people in India, 485 million in Nigeria, and more than 100 million each in Pakistan, Indonesia, and Sudan. Between 2 and 3.5 billion people will consequently find themselves living in climates that are fast becoming uninhabitable. As Professor Marten Scheffer, one of the lead authors of the study, put it, "Average temperatures over 29°C are unlivable. You'd have to move or adapt. But there are limits to adaptation. If you have enough money and energy, you can use air conditioning and fly in food and then you might be OK. But that is not the case for most people."[29] In other words, a significant segment of humanity—approximately 30 percent of the world's population—will have to move or die within the next fifty years, according to Scheffer's research.

The result will be crises that increasingly ramify across borders—but *mainly in regions of the Global South* rather than in the core imperial countries. The World Bank's September 2021 report *Groundswell: Preparing for Internal Climate Migration* states that climate change is particularly likely to impact regions already most afflicted by poverty and vulnerability. It predicts that by 2050, countries across sub-Saharan Africa could have as many as 85.7 million internally displaced people; East Asia and the Pacific, 48.4 million; South Asia, 40.5 million (with nearly 20 million in Bangladesh alone); North Africa and the Middle East, 19.3 million; Latin America and the Caribbean, 17.1 million; and Eastern Europe and Central Asia, 5.1 million.

It is no mystery who is responsible for this coming epic human convulsion. The world's ten largest historic emitters have produced 72 percent of total greenhouse gas emissions in the world dating back to 1850. Seven countries, according to UN climate agreements, bear particular responsibility as a result of their historic emissions and because of their wealth and levels of development. Grouped in a larger contingent of states that the UN calls "Annex II" countries, these seven countries are the United States, Germany, Japan, the United Kingdom, Canada, France, and Australia. Collectively, they are responsible for 48 percent of historic emissions. These core imperial nations have colonized the atmosphere and must make amends.

There is a strong case to be made for climate reparations, a principle acknowledged even within the UN Climate Process, which has historically been dominated precisely by the core imperial nations and affiliated corporate agendas. The 1992 UN Framework Convention on Climate Change recognizes the unjust burden the climate crisis puts on low-income countries through the principle known as "common but differentiated responsibilities" (Article 3). The CBDR principle recognizes that each country has a responsibility to reduce greenhouse gas emissions, but the responsibility must be differentiated in accordance with its social and economic conditions because not all countries contributed equally. In other words, the rich polluting countries have a responsibility to both reduce their emissions proportionally *more* than poorer countries, as well as to provide the finance to poor countries so they can leapfrog development based on fossil fuels, build renewable energy economies, and adapt to the impacts of climate change.

What would reparations mean for climate refugees? As we saw in the introduction, the Cochabamba Declaration articulated many of the terms that still offer a horizon for global ecological transformation. In addition to insisting that the wealthy nations decolonize the atmosphere by reducing and removing their emissions, the Cochabamba Declaration also sets out a framework for climate debt as it pertains to climate-induced migration. The activists gathered in Cochabamba argued that there are three specific forms of debt: development, adaptation, and migration debt. The first two are relatively uncontroversial pillars of climate negotiations. As a less familiar demand, the latter form of debt needs unpacking. The working group on climate migration made three central demands (emphasis mine):

1. We demand political, economic, social and cultural patterns, in which *the right to move and displace freely is respected*, also a pattern that respects the right not to migrate and not to be displaced by force;

2. We demand the *promotion of a human rights treaty for climate migrants*, recognized and applicable at a global scale, one of binding character and therefore claimable, so that climatic migrants have the same rights and obligations as the citizens of the country of destination;

3. We demand the *creation of an economic fund*, funded mainly by the countries in the center of capitalism and huge transnational corporations, that are held main responsible for climate change, destined to meet the needs of both internal and international climatic migrants.[30]

The Cochabamba Declaration incorporates these calls into a framework that challenges the juridical invisibility of climate refugees by calling for recognition and protection. Beyond that, though, it also calls for a sweeping transformation in the underlying cultural, economic, and political arrangements that have fostered the refugee interdiction and deportation regime. This is, in essence, a demand for border abolition. Lastly, the declaration insists that core imperial nations and transnational corporations create an economic fund that would help facilitate the internal and international forms of mobility generated by the rich nations' pollution of the global atmospheric commons.

The core imperial countries of the Global North have largely failed to meet these demands for the reparation of various forms of climate debt, including migration debt. In 2009, at the UN Climate Conference in Copenhagen, wealthy nations committed to mobilizing $100 billion a year in climate finance by 2020 for developing countries. Little justification was given for this relatively paltry sum. But the richest countries have fallen far short of even these inadequate promises every year since 2009. The latest figures produced by the Organization for Economic Co-operation and Development show that its member countries provided only around

$80 billion in climate finance in 2019, and their 2020 commitments offered only an additional $1.6 billion. Worse still, the nonprofit Oxfam International has tracked these wealthy-nation financial commitments and has noted that up to 80 percent of the finance comes in the form of loans rather than grants. These loans entrench injustice by adding to recipient countries' already ballooning debt burdens. And this in turn is a massive impediment to the efforts of these countries to adapt adequately to the intensifying climate crisis. Instead of building public schools and libraries that can serve as refuges during hurricane or typhoon season, poor countries are forced to pay off fat-cat bankers based in wealthy nations.

Rather than paying the climate debt they owe to Global South nations, rich countries have spent lavishly on walls, fences, cages, and paramilitary policing. In an important report on the border-security-industrial complex, researchers at the Transnational Institute (TNI) totaled the average yearly contribution to climate financing from the Annex II countries between 2013 and 2018, and compared these totals to their spending on border and immigration enforcement over the same period.[31] The combined annual border and immigration enforcement spending of the Annex II countries averages $33.1 billion. This sum is 2.3 times more than their climate financing. According to the TNI report, four of these wealthy countries have border and immigration enforcement budgets that are higher than their climate financing budgets. Canada leads this shameful crew, spending on average fifteen times more on border and immigration enforcement than on climate financing. The US, which spends more in absolute terms on border policing than any other nation—$19.6 billion in 2021—designated only $1.8 billion for climate financing. In other words, it spent nearly eleven times as much on border policing than it did on helping countries cope with the carbon emissions it has had such an outsize role in generating.

Sixty-three border walls have been built between nations over the last fifty years. Yet, as many scholars have documented in recent years, walls are only one element of border fortification.[32] Countries have spent lavishly on militarized border guards, and on increasingly deadly forms of surveillance technology to monitor borders. In addition, borders are no longer located exclusively at the geographical points of contact between nations.

The border-security apparatus increasingly penetrates into the cities and rural areas of wealthy nations, and is simultaneously extending outward into surrounding countries, which have been prodded in recent years into providing interdiction services for migrants traversing their territory.

The border-security-industrial complex has boomed over the last few decades. According to TNI researchers, between 2008 and 2020 the US federal government issued to private companies more than 105,000 contracts worth $55 billion. Pivotal border profiteering companies that researchers identified include CoreCivic, Deloitte, Elbit Systems, GEO Group, General Atomics, General Dynamics, G4S, IBM, Leidos, Lockheed Martin, L3Harris, Northrop Grumman, and Palantir. These companies provide private detention facilities, surveillance technology, biometric systems, databases, armored transportation, and drones, among other technologies of control and punishment.

It may not come as a surprise that many of these companies are key players in the US military-industrial complex. Perhaps more unexpected is the fact that many of the leading border policing companies have contracts to provide security for fossil fuel companies. In fact, in many cases the same rich people (or, in the vast majority of cases, wealthy white men) sit on the boards of fossil fuel and border-security companies. The border-security industry thus protects the corporations that are destroying the planet, while also detaining and punishing the people who are displaced by the impact of the carbon emissions for which these corporations are responsible.

The border-security-industrial complex is a key institutional form of the global right wing, one that germinates from and in turn nurtures the noxious ideologies of white supremacy and eco-fascism that are increasingly permeating the public sphere in wealthy nations. Fascism emerges from a capitalist world-system in crisis. Capitalism as a mode of production and a social system *requires* people to be destructive of the environment. Three destructive aspects of the capitalist system stand out when we consider the question of its ecological foundations: (1) capitalism must expand ceaselessly in order to survive; (2) it inherently tends to degrade the conditions of its own production; and (3) it generates a chaotic and competitive world-system, which in turn intensifies its ecological contradictions.[33]

As radical geographer David Harvey argues, capital is characterized by a "bad infinity": the system must expand constantly because it is grounded in profit, or what Marx calls the generation of surplus value.[34] It should be self-evident that an economy and a culture based on 3 percent compound growth will eventually annihilate the finite planetary ecosystem on which it is based. Movements for environmentalism from below thus inherently work toward the eradication of capitalism.

The movement of people is a form of climate adaptation and deserves support from climate justice movements in the Global North, whether those people ultimately arrive in core imperial nations or not. What would this look like in practical terms? Michael Gerrard, the director of the Sabin Center for Climate Change Law at Columbia University in New York, has suggested that "rather than leaving vast numbers of victims of a warmer world stranded, without any place allowing them in, industrialized countries ought to pledge to take on a share of the displaced population equal to how much each nation has historically contributed to emissions of the greenhouse gases that are causing this crisis."[35] According to the Climate Equity Reference Calculator, the US is currently the source of 30 percent of greenhouse gas emissions; the EU, 18 percent; China, 16 percent; Japan, 4 percent; Canada, 3 percent; and Australia, 2 percent. If climate-related migrants were admitted in the same proportion, for every 100 million the US would take in 30 million, the EU, 18 million, and so on.

Paying climate reparations by giving harbor to climate refugees in this manner is a good concrete demand, a kind of non-reformist reform. But ultimately, the insight that migration is a form of climate adaptation leads inescapably to a demand for the abolition of borders.[36] Nation-state borders emerged out of long histories of colonialism, imperialism, and white supremacy. They are a relatively recent invention, dating back to the Chinese Exclusion Act of 1882 in the US and similar legislation in other settler colonies. Today their existence continues to support a global order riven with inequality, injustice, and racist violence. Borders are machines for the production of a violent, colonial, and fascist global order. Razor-wire fences, armed men, surveillance drones, and all the other paraphernalia of the border-security-industrial complex do not stop people, they only make movement more perilous. And borders are not just dangerous for migrants.

The growth of a militarized border-security apparatus helps reinforce and feed authoritarian populism among segments of the public in core imperial nations, and in other countries to boot. This fascist creep must be stopped.

⌣

Borders do not prevent people from moving. Large numbers of people are already on the move in response to a climate crisis not of their own making. Given the crisis's trajectory, their numbers are only going to grow. Although there is no global organization of climate refugees, no real equivalent of La Via Campesina's transnational organizing of peasants, climate refugees are nonetheless a global movement (in both senses of the term). Yet people like Cruz Martinez and Ernest Francois, along with untold numbers of migrants and refugees, are fighting the racist border and deportation regime tenaciously. They are making history by reconfiguring human culture and political organization on a vast, transnational scale. Indeed, climate refugees could be called one of the most consequential movements in the history of humanity, and it seems likely that most people will ultimately become climate refugees, although under radically different conditions.

Today, with the barbaric character of the global capitalist system laid bare by the coronavirus crash, we more than ever need new narratives of possible worlds.[37] This is as much an issue of imagination as it is of design since the current crisis is ultimately a product of the capitalist foreclosure of emancipatory possibilities.[38] The climate crisis will surely bear down harder in coming years, and with it the social pressures that underscore the burning need for such alternative visions. As I suggested in the introductory chapter, there is a long history of solidarity between environmental and climate justice movements in Global North nations and similar movements in the Global South. Nowhere will such forms of international solidarity be more important than in the struggle against the border-security-industrial complex. Movements in the Global North are already fighting the border-security and deportation regime, not just for the sake of the peoples of the Global South but also to fight fascist movements in the Global North.

In the face of intensifying crises, environmentalism from below holds open the prospect of a future founded not on fear, militarization, and an eco-fascist mindset, but on conviviality, borderless mobility, and global solidarity. The feelings of solidarity that emanate from environmentalism from below surrounded me as I stood alongside friends old and new from New York City–based environmental justice movements at the closing ceremony of the World People's Conference on Climate Change and the Rights of Mother Earth. Behind us was a large and extremely vocal contingent of Argentinians from a group called Los Pibes (The Kids), waving bright blue banners in the early morning sunlight and singing radical songs that sounded like football chants. A procession of Indigenous peoples from across the Americas, enveloped by billowing clouds of incense, moved through the crowd, performing a ritual to beseech Pachamama, or Mother Earth, for forgiveness for our desecration of the planet. The ceremony took place on April 22, 2010, Mother Earth Day.

On that day, the thirty-two thousand climate justice activists who had gathered in Cochabamba collectively proclaimed the Universal Declaration of Rights of Mother Earth. This document recognizes that Earth is an indivisible, living community of interrelated and interdependent beings with inherent rights.[39] It also defines fundamental human responsibilities in relation to other beings and to the community as a whole. The declaration proclaimed our collective determination to replace exploitative values, worldviews, and political, economic, and legal systems that uphold them with those that respect and defend the rights and harmonious co-existence of all beings.

This transformed orientation is fundamental to the ongoing fight for a decolonial, anti-capitalist, and anti-patriarchal ecological reconstruction of the world. It is only fitting that a book dedicated to exploring and celebrating movements for environmentalism from below conclude with a reaffirmation of this collective vision of our world transformed.

# UNIVERSAL DECLARATION
# OF RIGHTS OF MOTHER EARTH

World People's Conference on Climate Change
and the Rights of Mother Earth

Cochabamba, Bolivia

April 22, 2010

## Preamble

We, the peoples and nations of Earth:

considering that we are all part of Mother Earth, an indivisible, living community of interrelated and interdependent beings with a common destiny;

gratefully acknowledging that Mother Earth is the source of life, nourishment and learning and provides everything we need to live well;

recognizing that the capitalist system and all forms of depredation, exploitation, abuse and contamination have caused great destruction, degradation and disruption of Mother Earth, putting life as we know it today at risk through phenomena such as climate change;

convinced that in an interdependent living community it is not possible to recognize the rights of only human beings without causing an imbalance within Mother Earth;

affirming that to guarantee human rights it is necessary to recognize and defend the rights of Mother Earth and all beings in her and that there are existing cultures, practices and laws that do so;

conscious of the urgency of taking decisive, collective action to transform structures and systems that cause climate change and other threats to Mother Earth;

proclaim this Universal Declaration of the Rights of Mother Earth, and call on the General Assembly of the United Nation to adopt it, as a common standard of achievement for all peoples and all nations of the world, and to the end that every individual and institution takes responsibility for promoting through teaching, education, and consciousness raising, respect for the rights recognized in this Declaration and ensure through prompt and progressive measures and mechanisms, national and international, their universal and effective recognition and observance among all peoples and States in the world.

## Article 1. Mother Earth

(1) Mother Earth is a living being.

(2) Mother Earth is a unique, indivisible, self-regulating community of interrelated beings that sustains, contains and reproduces all beings.

(3) Each being is defined by its relationships as an integral part of Mother Earth.

(4) The inherent rights of Mother Earth are inalienable in that they arise from the same source as existence.

(5) Mother Earth and all beings are entitled to all the inherent rights recognized in this Declaration without distinction of any kind, such as may be made between organic and inorganic beings, species, origin, use to human beings, or any other status.

(6) Just as human beings have human rights, all other beings also have

rights which are specific to their species or kind and appropriate for their role and function within the communities within which they exist.

(7) The rights of each being are limited by the rights of other beings and any conflict between their rights must be resolved in a way that maintains the integrity, balance and health of Mother Earth.

## Article 2. Inherent Rights of Mother Earth

(1) Mother Earth and all beings of which she is composed have the following inherent rights:

(a) the right to life and to exist;

(b) the right to be respected;

(c) the right to regenerate its bio-capacity and to continue its vital cycles and processes free from human disruptions;

(d) the right to maintain its identity and integrity as a distinct, self-regulating and interrelated being;

(e) the right to water as a source of life;

(f) the right to clean air;

(g) the right to integral health;

(h) the right to be free from contamination, pollution and toxic or radioactive waste;

(i) the right to not have its genetic structure modified or disrupted in a manner that threatens its integrity or vital and healthy functioning;

(j) the right to full and prompt restoration for violations of the rights recognized in this Declaration caused by human activities.

(2) Each being has the right to a place and to play its role in Mother Earth for her harmonious functioning.

(3) Every being has the right to well-being and to live free from torture or cruel treatment by human beings.

## Article 3. Obligations of Human Beings to Mother Earth

(1) Every human being is responsible for respecting and living in harmony with Mother Earth.

(2) Human beings, all States, and all public and private institutions must:

(a) act in accordance with the rights and obligations recognized in this Declaration;

(b) recognize and promote the full implementation and enforcement of the rights and obligations recognized in this Declaration;

(c) promote and participate in learning, analysis, interpretation and communication about how to live in harmony with Mother Earth in accordance with this Declaration;

(d) ensure that the pursuit of human well-being contributes to the well-being of Mother Earth, now and in the future;

(e) establish and apply effective norms and laws for the defence, protection and conservation of the rights of Mother Earth;

(f) respect, protect, conserve and where necessary, restore the integrity, of the vital ecological cycles, processes and balances of Mother Earth;

(g) guarantee that the damages caused by human violations of the inherent rights recognized in this Declaration are rectified and that those responsible are held accountable for restoring the integrity and health of Mother Earth;

(h) empower human beings and institutions to defend the rights of Mother Earth and of all beings;

(i) establish precautionary and restrictive measures to prevent human activities from causing species extinction, the destruction of ecosystems or the disruption of ecological cycles;

(j) guarantee peace and eliminate nuclear, chemical and biological weapons;

(k) promote and support practices of respect for Mother Earth and all beings, in accordance with their own cultures, traditions and customs;

(l) promote economic systems that are in harmony with Mother Earth and in accordance with the rights recognized in this Declaration.

## Article 4. Definitions

(1) The term "being" includes ecosystems, natural communities, species and all other natural entities which exist as part of Mother Earth.

(2) Nothing in this Declaration restricts the recognition of other inherent rights of all beings or specified beings.

# PEOPLE'S AGREEMENT

# OF COCHABAMBA

World People's Conference on Climate Change
and the Rights of Mother Earth

April 22nd, Cochabamba, Bolivia

## PEOPLE'S AGREEMENT

Today, our Mother Earth is wounded and the future of humanity is in danger.

If global warming increases by more than 2 degrees Celsius, a situation that the "Copenhagen Accord" could lead to, there is a 50% probability that the damages caused to our Mother Earth will be completely irreversible. Between 20% and 30% of species would be in danger of disappearing. Large extensions of forest would be affected, droughts and floods would affect different regions of the planet, deserts would expand, and the melting of the polar ice caps and the glaciers in the Andes and Himalayas would worsen. Many island states would disappear, and Africa would suffer an increase in temperature of more than 3 degrees Celsius. Likewise, the production of food would diminish in the world, causing catastrophic impact on the survival of inhabitants from vast regions in the planet, and the number of people in the world suffering from hunger would increase dramatically, a figure that already exceeds 1.02 billion people. The corporations and governments of the so-called "developed" countries, in complicity with a segment of the scientific community, have led us to discuss climate change as a problem limited to the rise in temperature without questioning the cause, which is the capitalist system.

We confront the terminal crisis of a civilizing model that is patriarchal and based on the submission and destruction of human beings and nature that accelerated since the industrial revolution.

The capitalist system has imposed on us a logic of competition, progress and limitless growth. This regime of production and consumption seeks profit without limits, separating human beings from nature and imposing a logic of domination upon nature, transforming everything into commodities: water, earth, the human genome, ancestral cultures, biodiversity, justice, ethics, the rights of peoples, and life itself.

Under capitalism, Mother Earth is converted into a source of raw materials, and human beings into consumers and a means of production, into people that are seen as valuable only for what they own, and not for what they are.

Capitalism requires a powerful military industry for its processes of accumulation and imposition of control over territories and natural resources, suppressing the resistance of the peoples. It is an imperialist system of colonization of the planet.

Humanity confronts a great dilemma: to continue on the path of capitalism, depredation, and death, or to choose the path of harmony with nature and respect for life.

It is imperative that we forge a new system that restores harmony with nature and among human beings. And in order for there to be balance with nature, there must first be equity among human beings. We propose to the peoples of the world the recovery, revalorization, and strengthening of the knowledge, wisdom, and ancestral practices of Indigenous Peoples, which are affirmed in the thought and practices of "Living Well," recognizing Mother Earth as a living being with which we have an indivisible, interdependent, complementary and spiritual relationship. To face climate change, we must recognize Mother Earth as the source of life and forge a new system based on the principles of:

- harmony and balance among all and with all things;

- complementarity, solidarity, and equality;

- collective well-being and the satisfaction of the basic necessities of all;

- people in harmony with nature;

- recognition of human beings for what they are, not what they own;

- elimination of all forms of colonialism, imperialism and interventionism;

- peace among the peoples and with Mother Earth.

The model we support is not a model of limitless and destructive development. All countries need to produce the goods and services necessary to satisfy the fundamental needs of their populations, but by no means can they continue to follow the path of development that has led the richest countries to have an ecological footprint five times bigger than what the planet is able to support. Currently, the regenerative capacity of the planet has been already exceeded by more than 30 percent. If this pace of over-exploitation of our Mother Earth continues, we will need two planets by the year 2030. In an interdependent system in which human beings are only one component, it is not possible to recognize rights only to the human part without provoking an imbalance in the system as a whole. To guarantee human rights and to restore harmony with nature, it is necessary to effectively recognize and apply the rights of Mother Earth. For this purpose, we propose the attached project for the Universal Declaration on the Rights of Mother Earth, in which it's recorded that:

- The right to live and to exist;

- The right to be respected;

- The right to regenerate its bio-capacity and to continue its vital cycles and processes free of human alteration;

- The right to maintain their identity and integrity as differentiated beings, self-regulated and interrelated;

- The right to water as the source of life;

- The right to clean air;

- The right to comprehensive health;

- The right to be free of contamination and pollution, free of toxic and radioactive waste;

- The right to be free of alterations or modifications of its genetic structure in a manner that threatens its integrity or vital and healthy functioning;

- The right to prompt and full restoration for violations to the rights acknowledged in this Declaration caused by human activities.

The "shared vision" seeks to stabilize the concentrations of greenhouse gases to make effective the Article 2 of the United Nations Framework Convention on Climate Change, which states that "the stabilization of greenhouse gases concentrations in the atmosphere to a level that prevents dangerous anthropogenic inferences for the climate system." Our vision is based on the principle of historical common but differentiated responsibilities, to demand the developed countries to commit with quantifiable goals of emission reduction that will allow to return the concentrations of greenhouse gases to 300 ppm, therefore the increase in the average world temperature to a maximum of one degree Celsius.

Emphasizing the need for urgent action to achieve this vision, and with the support of peoples, movements and countries, developed countries should commit to ambitious targets for reducing emissions that permit the achievement of short-term objectives, while maintaining our vision in favor of balance in the Earth's climate system, in agreement with the ultimate objective of the Convention.

The "shared vision for long-term cooperative action" in climate change negotiations should not be reduced to defining the limit on temperature increases and the concentration of greenhouse gases in the atmosphere, but must also incorporate in a balanced and integral manner measures regarding capacity building, production and consumption patterns, and other essential factors such as the acknowledging of the Rights of Mother Earth to establish harmony with nature.

Developed countries, as the main cause of climate change, in assuming their historical responsibility, must recognize and honor their climate debt in all of its dimensions as the basis for a just, effective, and scientific solution to climate change. In this context, we demand that developed countries:

- Restore to developing countries the atmospheric space that is occupied by their greenhouse gas emissions. This implies the decolonization of the atmosphere through the reduction and absorption of their emissions;

- Assume the costs and technology transfer needs of developing countries arising from the loss of development opportunities due to living in a restricted atmospheric space;

- Assume responsibility for the hundreds of millions of people that will be forced to migrate due to the climate change caused by these countries, and eliminate their restrictive immigration policies, offering migrants a decent life with full human rights guarantees in their countries;

- Assume adaptation debt related to the impacts of climate change on developing countries by providing the means to prevent, minimize, and deal with damages arising from their excessive emissions;

- Honor these debts as part of a broader debt to Mother Earth by adopting and implementing the United Nations Universal Declaration on the Rights of Mother Earth.

The focus must not be only on financial compensation, but also on restorative justice, understood as the restitution of integrity to our Mother Earth and all its beings.

We deplore attempts by countries to annul the Kyoto Protocol, which is the sole legally binding instrument specific to the reduction of greenhouse gas emissions by developed countries.

We inform the world that, despite their obligation to reduce emissions, developed countries have increased their emissions by 11.2% in the period from 1990 to 2007.

During that same period, due to unbridled consumption, the United States of America has increased its greenhouse gas emissions by 16.8%, reaching an average of 20 to 23 tons of $CO_2$ per person. This represents 9 times more than that of the average inhabitant of the "Third World," and 20 times more than that of the average inhabitant of Sub-Saharan Africa.

We categorically reject the illegitimate "Copenhagen Accord" that allows developed countries to offer insufficient reductions in greenhouse gases based in voluntary and individual commitments, violating the environmental integrity of Mother Earth and leading us toward an increase in global temperatures of around 4°C.

The next Conference on Climate Change to be held at the end of 2010 in Mexico should approve an amendment to the Kyoto Protocol for the second commitment period from 2013 to 2017 under which developed countries must agree to significant domestic emissions reductions of at least 50% based on 1990 levels, excluding carbon markets or other offset mechanisms that mask the failure of actual reductions in greenhouse gas emissions.

We require first of all the establishment of a goal for the group of developed countries to achieve the assignment of individual commitments for each developed country under the framework of complementary efforts among each one, maintaining in this way Kyoto Protocol as the route to emissions reductions.

The United States, as the only Annex 1 country on Earth that did not ratify the Kyoto Protocol, has a significant responsibility toward all peoples of the world to ratify this document and commit itself to respecting and complying with emissions reduction targets on a scale appropriate to the total size of its economy.

We the peoples have the equal right to be protected from the adverse effects of climate change and reject the notion of adaptation to climate change as understood as a resignation to impacts provoked by the historical emissions of developed countries, which themselves must adapt their modes of life and consumption in the face of this global emergency. We see it as imperative to confront the adverse effects of climate change, and consider adaptation to be a process rather than an imposition, as well as a tool that can serve to help offset those effects, demonstrating that it is possible to achieve harmony with nature under a different model for living.

It is necessary to construct an Adaptation Fund exclusively for addressing climate change as part of a financial mechanism that is managed in a sovereign, transparent, and equitable manner for all States. This

Fund should assess the impacts and costs of climate change in developing countries and needs deriving from these impacts, and monitor support on the part of developed countries. It should also include a mechanism for compensation for current and future damages, loss of opportunities due to extreme and gradual climactic events, and additional costs that could present themselves if our planet surpasses ecological thresholds, such as those impacts that present obstacles to "Living Well."

The "Copenhagen Accord" imposed on developing countries by a few States, beyond simply offering insufficient resources, attempts as well to divide and create confrontation between peoples and to extort developing countries by placing conditions on access to adaptation and mitigation resources. We also assert as unacceptable the attempt in processes of international negotiation to classify developing countries for their vulnerability to climate change, generating disputes, inequalities and segregation among them.

The immense challenge humanity faces of stopping global warming and cooling the planet can only be achieved through a profound shift in agricultural practices toward the sustainable model of production used by indigenous and rural farming peoples, as well as other ancestral models and practices that contribute to solving the problem of agriculture and food sovereignty. This is understood as the right of peoples to control their own seeds, lands, water, and food production, thereby guaranteeing, through forms of production that are in harmony with Mother Earth and appropriate to local cultural contexts, access to sufficient, varied and nutritious foods in complementarity with Mother Earth and deepening the autonomous (participatory, communal and shared) production of every nation and people.

Climate change is now producing profound impacts on agriculture and the ways of life of indigenous peoples and farmers throughout the world, and these impacts will worsen in the future.

Agribusiness, through its social, economic, and cultural model of global capitalist production and its logic of producing food for the market and not to fulfill the right to proper nutrition, is one of the principal causes of climate change. Its technological, commercial, and political approach only serves to deepen the climate change crisis and increase hunger in the world. For this reason, we reject Free Trade Agreements and Association Agreements and all forms of the application of Intellectual Property Rights to life, current technological packages (agrochemicals, genetic modification) and those that offer false solutions (biofuels, geo-engineering, nanotechnology, etc.) that only exacerbate the current crisis.

We similarly denounce the way in which the capitalist model imposes mega-infrastructure projects and invades territories with extractive projects, water privatization, and militarized territories, expelling indigenous peoples from their lands, inhibiting food sovereignty and deepening socio-environmental crisis.

We demand recognition of the right of all peoples, living beings, and Mother Earth to have access to water, and we support the proposal of the Government of Bolivia to recognize water as a Fundamental Human Right.

The definition of forests used in the negotiations of the United Nations Framework Convention on Climate Change, which includes plantations, is unacceptable. Monoculture plantations are not forests. Therefore, we require a definition for negotiation purposes that recognizes the native forests, jungles and the diverse ecosystems on Earth.

The United Nations Declaration on the Rights of Indigenous Peoples must be fully recognized, implemented and integrated in climate change negotiations. The best strategy and action to avoid deforestation and degradation and protect native forests and jungles is to recognize and guarantee collective rights to lands and territories, especially considering that most of the forests are located within the territories of indigenous peoples and nations and other traditional communities.

We condemn market mechanisms such as REDD (Reducing Emissions from Deforestation and Forest Degradation) and its versions + and + +, which are violating the sovereignty of peoples and their right to prior free and informed consent as well as the sovereignty of national States, the customs of Peoples, and the Rights of Nature.

Polluting countries have an obligation to carry out direct transfers of the economic and technological resources needed to pay for the restoration and maintenance of forests in favor of the peoples and indigenous ancestral organic structures. Compensation must be direct and in addition to the sources of funding promised by developed countries outside of the carbon market, and never serve as carbon offsets. We demand that countries stop actions on local forests based on market mechanisms and propose non-existent and conditional results. We call on governments to create a global program to restore native forests and jungles, managed and administered by the peoples, implementing forest seeds, fruit trees, and native flora. Governments should eliminate forest concessions and support the conservation of petroleum deposits in the ground and urgently stop the exploitation of hydrocarbons in forestlands.

We call upon States to recognize, respect and guarantee the effective implementation of international human rights standards and the rights of indigenous peoples, including the United Nations Declaration on the Rights of Indigenous Peoples under ILO Convention 169, among other relevant instruments in the negotiations, policies and measures used to meet the challenges posed by climate change. In particular, we call upon States to give legal recognition to claims over territories, lands and natural resources to enable and strengthen our traditional ways of life and contribute effectively to solving climate change.

We demand the full and effective implementation of the right to consultation, participation and prior, free and informed consent of indigenous peoples in all negotiation processes, and in the design and implementation of measures related to climate change.

Environmental degradation and climate change are currently reaching critical levels, and one of the main consequences of this is domestic and international migration. According to projections, there were already about 25 million climate migrants by 1995. Current estimates are around 50 million, and projections suggest that between 200 million and 1 billion people will become displaced by situations resulting from climate change by the year 2050.

Developed countries should assume responsibility for climate migrants, welcoming them into their territories and recognizing their fundamental rights through the signing of international conventions that provide for the definition of climate migrant and require all States to abide by determinations.

Establish an International Tribunal of Conscience to denounce, make visible, document, judge and punish violations of the rights of migrants, refugees and displaced persons within countries of origin, transit and destination, clearly identifying the responsibilities of States, companies and other agents.

Current funding directed toward developing countries for climate change and the proposal of the Copenhagen Accord are insignificant. In addition to Official Development Assistance and public sources, developed countries must commit to a new annual funding of at least 6% of GDP to tackle climate change in developing countries. This is viable considering that a similar amount is spent on national defense, and that 5 times more have been put forth to rescue failing banks and speculators, which raises serious questions about global priorities and political will. This funding should be direct and free of conditions, and should not interfere with the national sovereignty or self-determination of the most affected communities and groups.

In view of the inefficiency of the current mechanism, a new funding mechanism should be established at the 2010 Climate Change Conference in Mexico, functioning under the authority of the Conference of the Parties (COP) under the United Nations Framework Convention on Climate Change and held accountable to it, with significant representation of developing countries, to ensure compliance with the funding commitments of Annex 1 countries.

It has been stated that developed countries significantly increased their emissions in the period from 1990 to 2007, despite having stated that the reduction would be substantially supported by market mechanisms.

The carbon market has become a lucrative business, commodifying our Mother Earth. It is therefore not an alternative for tackl[ing] climate change, as it loots and ravages the land, water, and even life itself.

The recent financial crisis has demonstrated that the market is incapable of regulating the financial system, which is fragile and uncertain due to speculation and the emergence of intermediary brokers. Therefore, it would be totally irresponsible to leave in their hands the care and protection of human existence and of our Mother Earth.

We consider inadmissible that current negotiations propose the creation of new mechanisms that extend and promote the carbon market, for existing mechanisms have not resolved the problem of climate change nor led to real and direct actions to reduce greenhouse gases. It is necessary to demand fulfillment of the commitments assumed by developed countries under the United Nations Framework Convention on Climate Change regarding development and technology transfer, and to reject the "technology showcase" proposed by developed countries that only markets technology. It is essential to establish guidelines in order to create a multilateral and multidisciplinary mechanism for participatory control, management, and evaluation of the exchange of technologies. These technologies must be useful, clean and socially sound. Likewise, it is fundamental to establish a fund for the financing and inventory of technologies that are appropriate and free of intellectual property rights. Patents, in particular, should move from the hands of private monopolies to the public domain in order to promote accessibility and low costs.

Knowledge is universal, and should for no reason be the object of private property or private use, nor should its application in the form of technology. Developed countries have a responsibility to share their technology with developing countries, to build research centers in developing countries for the creation of technologies and innovations, and defend and promote their development and application for "living well." The world must recover and re-learn ancestral principles and approaches from native peoples to stop the destruction of the planet, as well as promote ancestral practices, knowledge and spirituality to recuperate the capacity for "living well" in harmony with Mother Earth.

Considering the lack of political will on the part of developed countries to effectively comply with commitments and obligations assumed under the United Nations Framework Convention on Climate Change and the Kyoto Protocol, and given the lack of a legal international organism to guard against and sanction climate and environmental crimes that violate the Rights of Mother Earth and humanity, we demand the creation of an International Climate and Environmental Justice Tribunal that has the legal capacity to prevent, judge and penalize States, industries and people that by commission or omission contaminate and provoke climate change.

Supporting States that present claims at the International Climate and Environmental Justice Tribunal against developed countries that fail to comply with commitments under the United Nations Framework Convention on Climate Change and the Kyoto Protocol including commitments to reduce greenhouse gases.

We urge peoples to propose and promote deep reform within the United Nations, so that all member States comply with the decisions of the International Climate and Environmental Justice Tribunal.

The future of humanity is in danger, and we cannot allow a group of leaders from developed countries to decide for all countries as they tried unsuccessfully to do at the Conference of the Parties in Copenhagen. This decision concerns us all. Thus, it is essential to carry out a global referendum or popular consultation on climate change in which all are consulted regarding the following issues: the level of emission reductions on the part of developed countries and transnational corporations, financing to be offered by developed countries, the creation of an International Climate Justice Tribunal, the need for a Universal Declaration of the Rights of Mother Earth, and the need to change the current capitalist system. The process of a global referendum or popular consultation will depend on process of preparation that ensures the successful development of the same.

In order to coordinate our international action and implement the results of this "Accord of the Peoples," we call for the building of a Global People's Movement for Mother Earth, which should be based on the principles of complementarity and respect for the diversity of origin and visions among its members, constituting a broad and democratic space for coordination and joint worldwide actions.

To this end, we adopt the attached global plan of action so that in Mexico, the developed countries listed in Annex 1 respect the existing legal framework and reduce their greenhouse gases emissions by 50%, and that the different proposals contained in this Agreement are adopted.

Finally, we agree to undertake a Second World People's Conference on Climate Change and the Rights of Mother Earth in 2011 as part of this process of building the Global People's Movement for Mother Earth and reacting to the outcomes of the Climate Change Conference to be held at the end of this year in Cancun, Mexico.

# ACKNOWLEDGMENTS

This book would literally have been inconceivable without the many fierce climate justice activists who converged in Cochabamba for the World People's Conference on Climate Change and the Rights of Mother Earth. As I hope I make clear in the introduction, attending this conference was one of the defining moments of my life: it made clear the dramatic stakes of the fight for climate justice, it laid out many of the cardinal principles and policies of the struggle, and it exposed me to some of the bravest and most brilliant activists I've ever met. I will forever be indebted to all of you.

In particular, I am grateful to the members of the North American delegation in Cochabamba, including Taleigh Bicicleta, Tanya Fields, Mychal Johnson, Ryan Mann-Hamilton, Byron Santiago, and many others. I count myself incredibly fortunate to have been your comrade in Cochabamba, and look forward to working with you in more struggles.

The movement that manifested in Cochabamba had of course been around much longer. Indeed, many of the people I met there had been active already for decades, and were quite clear about being the inheritors of centuries of struggle against colonialism and capitalism. And, of course, the movement kept going after Cochabamba. I was fortunate to be around movement activists in subsequent meetings, including the counter-summit held during COP17 in Durban, South Africa. I learned an immense amount during this counter-summit from organizations like La Via Campesina and the Indigenous Environmental Network. I am particularly grateful for the intellectual inspiration and personal warmth shared with me in Durban by Eddie Yuen, Patrick Bond, Nick Buxton, and Joel Kovel.

I also could not have written this book without the many generous people who agreed to speak with me during my research. Among those I count myself extremely fortunate to have talked to are Kolya Abramsky, Max Ajl, Skylar Bissom, Nick Buxton, Pedro Gadanho, Justin McGuirk, Lenore Manderson, David Montgomery, Pat Mooney, Greg Muttitt, Trevor Ngwane, Brototi Roy, Miguel Robles, Nader Tehrani, Peter Rosset, Sean Sweeney, and Jo Woodman. I am very grateful to you all for sharing your insights and perspectives on the struggle for climate justice. To you and to the many other people who discussed the ideas that appear in this book, I give my deepest thanks. Most of this book was written during the COVID pandemic, a trying period all around the world but particularly so for people in the Global South. I am deeply aware of the tough times during which you agreed to share time and analysis with me.

I want to thank my students and colleagues at the City University of New York (CUNY). Those of you at both the Graduate Center and the College of Staten Island helped provide the stimulating conversations and thoughtful questions that form an important part of the background for this work. Thanks, in particular, to the students in my Graduate Center seminar on the Commons and Commoning for animated discussions about some of the ideas that appear in this book: Robert Balun, Coline Chevrin, Diane Enobabor, Val Fryer-Davis, Zoe Goldstein, Charlie Markbreiter, Rafael Munia, Natasha Ochshorn, Judah Rubin, Flora De Tournay-Oden.

I also would like to thank CUNY and my union, the Professional Staff Congress, for the sabbatical time that allowed me to complete the book. Although CUNY is beleaguered in multiple ways, it is among the best examples in the US of a public institution dedicated to the greater good; it would not be so without the struggles of generations of union members.

I am immensely grateful to my CUNY research assistant Rebecca Teich for whipping my citations into shape, which was a tremendous help getting the manuscript across the finish line.

My friend Yates Mckee, who collaborated with me in the CUNY Climate Action Lab (CAL), has been a great inspiration to me both during the period of our collaboration in CAL and, more broadly, through his activist work with Decolonize This Place. I am also indebted to the crack cadre of activists in the Public Power NY campaign, with whom I worked during the time that I was completing this project: your dedication, political acumen, and will to win democratic renewable energy in New York has been an unending source of inspiration.

The book benefited from the time for research and writing made available by a grant provided by FORMAS (Swedish Research Council for Sustainable Development) under the National Research Programme on Climate (Contract: 2017 -01962_3). I would like to thank my friend and comrade Marco Armiero of the Environmental Humanities Lab in Stockholm for proposing the collaboration that resulted in this grant.

I'd like to thank my editor Anthony Arnove, who gave me immensely helpful suggestions not just about specific political points in the book but also about how to hone my writing to be as engaging and accessible as possible. Thanks, also, to Haymarket senior editor Katy O'Donnell for your engagement with my work, and to all the rest of the Haymarket crew for your support.

Last of all, I owe immense thanks to my family. Ann, Nigel, and Ginny, for the wonderful foundation you provided for my life. Sofia, for your care and love during difficult years. And Manijeh and Sholeh, not just for your quotidian love but for your deep and abiding dedication to justice and women's empowerment. You make me understand what the fight for a better world looks like every day.

# BIBLIOGRAPHY

Abani, Chris. *Graceland*. New York: Farrar, Straus and Giroux, 2004.

Abourahme, Nasser. "Of Monsters and Boomerangs: Colonial Returns in the Late Liberal City." *City* 22, no. 1 (March 2018): 106–15. https://doi.org/10.1080/13604813.2018.1434296.

Achtenberg, Emily. "Contested Development: The Geopolitics of Bolivia's TIPNIS Conflict." *NACLA Report on the Americas* 46, no. 2 (2013): 6–11. https://doi.org/ 10.1080/10714839.2013.11721987.

Adam, David. "US Planning to Weaken Copenhagen Climate Deal, Europe Warns." *The Guardian*, September 15, 2009.

Ahmad, Omair. "Climate Crisis Is Foundation of Indian Farmers' Protests." The Third Pole, January 25, 2021. https://www.thethirdpole.net/en/food/climate-crisis-is-foundation-of-indian-farmers-protests/.

Ahmed, Nafeez. "We Don't Mine Enough Rare Earth Metals to Replace Fossil Fuels with Renewable Energy." *Vice*, December 12, 2018. https://www.vice.com/en/article/a3mavb/we-dont-mine-enough-rare-earth-metals-to-replace-fossil-fuels-with-renewable-energy.

Al Jazeera Staff. "India Unveils Renewable Energy Ambitions with Big Solar Push." Al Jazeera, November 3, 2021. https://www.aljazeera.com/gallery/2021/11/3/india-solar-renewable-energy-electricity-climate-crisis.

Alier, Juan Martínez. *The Environmentalism of the Poor: A Study of Ecological Conflicts and Valuation*. Northampton, MA: Edward Elgar, 2002.

Almond, R E A, M. Grooten, and T. Petersen, eds. *Living Planet Report 2020: Bending the Curve of Biodiversity Loss*. Gland, Switzerland: WWF, 2020. https://www.worldwildlife.org/publications/living-planet-report-2020.

Alternative Information and Development Centre. "Migrant Marikana: The Shifts in the Migrant Labour System." AIDC, August 23, 2017. https://aidc.org.za/migrant-marikana-shifts-migrant-labour-system/.

———. "Their Just Transition and Ours." AIDC, December 9, 2019. https://aidc.org.za/their-just-transition-and-ours/.

Amnesty International India. *"When Land Is Lost, Do We Eat Coal?"* Coal Min-

*ing and Violations of Adivasi Rights in India.* Amnesty International India, July 2016. https://www.amnesty.org/en/documents/asa20/4391/2016/en/.

Anghie, Antony. "Inequality, Human Rights, and the New International Economic Order." *Humanity Journal* 10, no. 3 (February 2020). http://humanityjournal.org/issue-10-3/.

Arboleda, Martín. *Planetary Mine: Territories of Extraction under Late Capitalism.* London: Verso Books, 2020.

Arneil, Barbara. *John Locke and America: The Defense of English Colonialism.* New York: Oxford University Press, 1998.

Aronoff, Kate. "The Socialist Win in Bolivia and the New Era of Lithium Extraction." *New Republic*, October 19, 2019. https://newrepublic.com/article/159848/socialist-win-bolivia-new-era-lithium-extraction.

Aronoff, Kate, et al. *A Planet to Win: Why We Need a Green New Deal.* New York: Verso Books, 2019.

Arsenault, Chris. "Only 60 Years of Farming Left If Soil Degradation Continues." *Scientific American*, December 5, 2014. https://www.scientificamerican.com/article/only-60-years-of-farming-left-if-soil-degradation-continues/#:~:text=ROME%20(Thomson%20Reuters%20Foundation)%20%2D,UN%20official%20said%20on%20Friday.

Ashley, Brian, et al. *One Million Climate Jobs: Moving South Africa Forward on a Low-Carbon, Wage-Led, and Sustainable Path.* Alternative Information and Development Centre, December 2016.

Ashman, Sam. "SA's Climate Crisis Is Embedded in Coal and Exports." *New Frame*, August 30, 2021. https://www.newframe.com/sas-climate-crisis-is-embedded-in-coal-and-exports/.

Ashman, Sam, Ben Fine, and Susan Newman. "The Crisis in South Africa: Neoliberalism, Financialization and Uneven and Combined Development." *Socialist Register* 47 (2011): 174–95.

Ashman, Sam, Seeraj Mohamed, and Susan Newman. "The Financialisation of the South African Economy and Its Impact on Economic Growth and Employment." *Development* 59, no. 1 (2013). http://doi.org/10.1057/s41301-017-0065-1.

Asian Centre for Human Rights. *The State of Encounter Killings in India: Target, Detain, Torture, Execute.* Asian Centre for Human Rights, November 2018. https://www.ecoi.net/en/document/1457651.html.

Athanasiou, Tom. "Only a Global Green New Deal Can Save the Planet." *The Nation*, September 18, 2019. https://www.thenation.com/article/archive/green-new-deal-sanders/.

Ayamolowo, Oladimeji Joseph, P. T. Manditereza, and K. Kusakana. "South Africa Power Reforms: The Path to a Dominant Renewable Energy-Sourced

Grid." *Energy Reports* 8 (April 2022): 1208–15. https://doi.org/10.1016/j.egyr.2021.11.100.

Ayma, Evo Morales, "We in the Social Movements Know the Problems and Also the Solutions." In *The Earth Does Not Belong to Us, We Belong to the Earth*. Bolivian Ministry of Foreign Relations Publication, 2010.

Bade, Gavin. "Power to the People: Bernie Calls for Federal Takeover of Electricity Production." *Politico*, February 2, 2020. https://www.politico.com/news/2020/02/02/bernie-sanders-climate-federal-electricity-production-110117.

Baigrie, Bruce. "Eskom, Unbundling, and Decarbonization." *Phenomenal World* (blog), February 14, 2022. https://www.phenomenalworld.org/analysis/eskom-unbundling-and-decarbonization/.

Baiocchi, Gianpaolo. *We, the Sovereign*. Medford, MA: Polity Press, 2018.

Balaban, Utku. "The Enclosure of Urban Space and Consolidation of the Capitalist Land Regime in Turkish Cities." *Urban Studies* 48, no. 10 (December 2010): 2162–79 https://doi.org/10.1177/004209801038095.

Barbier, Edward B. *A Global Green New Deal: Rethinking the Economic Recovery*. United Nations Environmental Program, 2009.

Barrington-Leigh, Christopher, and Adam Millard-Ball. "Global Trends toward Urban Street-Network Sprawl." *Proceedings of the National Academy of Sciences* 117, no. 4 (January 2020): 1941–50.

Barton, Gregory. *Empire Forestry and the Origins of Environmentalism*. New York: Cambridge University Press, 2002.

Baskin, Jeremy. *Striking Back: A History of Cosatu*. New York: Verso Books, 1991.

Bastani, Aaron. *Fully Automated Luxury Communism: A Manifesto*. New York: Verso Books, 2019.

Beard, Victoria A., Anjali Mahendra, and Michael I. Westphal. "Towards a More Equal City: Framing the Challenges and Opportunities." World Resources Institute, October 13, 2016. https://www.wri.org/research/towards-more-equal-city-framing-challenges-and-opportunities.

Beauchamp, Zack. "Why Bernie Sanders Failed." *Vox*, April 10, 2020. https://www.vox.com/policy-and-politics/2020/4/10/21214970/bernie-sanders-2020-lost-class-socialism.

Beckett, Andy. "Accelerationism: How a Fringe Philosophy Predicted the Future We Live In." *The Guardian*, May 11, 2017.

Berlant, Lauren. *Cruel Optimism*. Durham, NC: Duke University Press, 2012.

Biel, Robert. *Sustainable Food Systems: The Role of the City*. London: UCL Press, 2016.

BizNews. "How to Steal a Billion from Eskom—and Leave SA in Darkness." *BizNews*, April 20, 2022. https://www.biznews.com/energy/2022/04/20/

losing-power-steal-eskom-darkness.

Bogost, Ian. "Can You Sue a Robocar?" *The Atlantic*, March 20, 2018.

Bomnalli, Manjunath. "Coal India Will Not Be Privatised." *Deccan Herald*, July 10, 2020. https://www.deccanherald.com/state/karnataka-districts/coal-in-dia-will-not-be-privatised-given-target-of-billion-tonnes-output-by-2023-pralhad-joshi-859390.html.

Boundja, Patrick. "Wild Places Bateke Plateaux Landscape." WCS Congo, 2018. https://congo.wcs.org/wild-places/bateke-plateaux.aspx.

Brand, Ulrich, and Markus Wissen. "The Imperial Mode of Living." In *The Routledge Handbook of Ecological Economics: Nature and Society*, edited by Clive L. Spash, 152–61. New York: Routledge, 2017.

Buchner, Barbara, et al. "Global Landscape of Climate Finance 2019." Climate Policy Initiative, November 7, 2019. https://www.climatepolicyinitiative.org/publication/global-landscape-of-climate-finance-2019/.

Building Bridges Collective. *Space for Movement?: Reflections from Bolivia on Climate Justice, Social Movements and the State*. Leeds, UK: Footprint Workers Co-op, 2010.

Bulmer, Ruppert, et al. *Global Perspective on Coal Jobs and Managing Labor Transition Out of Coal: Key Issues and Policy Responses*. Washington, DC: World Bank, December 2021. https://openknowledge.worldbank.org/handle/10986/37118.

Burnham, Philip. *Indian Country, God's Country: Native Americans and the National Parks*. Washington, DC: Island Press, 2000.

Busby, Joshua W., et al. "The Case for US Cooperation with India on a Just Transition Away from Coal." *Brookings* (blog), April 20, 2021. https://www.brookings.edu/research/the-case-for-us-cooperation-with-india-on-a-just-transition-away-from-coal/.

Büscher, Bram, and Robert Fletcher, *The Conservation Revolution: Radical Ideas for Saving Nature beyond the Anthropocene*. New York: Verso Books, 2020.

Buxton, Edward. *Two African Trips, with Notes and Suggestions on Big Game Preservation in Africa*. London: E. Stanford, 1902.

Buxton, Nick, and Ben Hayes. *The Secure and the Dispossessed: How the Military and Corporates Are Shaping a Climate-Changed World*. London: Pluto, 2015.

Cardoso, Andrea, and Ethemcan Turhan. "Examining New Geographies of Coal: Dissenting Energyscapes in Colombia and Turkey." *Applied Energy* 224 (August 2018): 398–408. https://doi.org/10.1016/j.apenergy.2018.04.096.

CARE Food and Water Systems. *Left Out and Left Behind: Ignoring Women Will Prevent Us from Solving the Hunger Crisis*. November 2020.

Cassey, Brian. "India's Ancient Tribes Battle to Save Their Forest Home from Mining." *The Guardian*, February 10, 2020.

Centre for Development and Enterprise. "VIEWPOINTS | Reviving a De-
    clining Mining Industry." *CDE – The Centre for Development and Enterprise*
    (blog), August 4, 2020. https://www.cde.org.za/viewpoints-reviving-a-de-
    clining-mining-industry/.

Chait, Jonathan. "Obama Had a Green New Deal, and It Worked. Let's Do
    That Again." *New York Magazine*, April 26, 2019. https://nymag.com/intel-
    ligencer/2019/04/obamas-green-new-deal-worked-climate-change.html.

Chakravartty, Anupam. "Latest Kaziranga Expansion Brings Back Fear of
    Evictions among Residents." *Mongabay*, November 25, 2020. https://india.
    mongabay.com/2020/11/latest-kaziranga-expansion-brings-back-fear-of-
    evictions-among-residents/.

Chamoiseau, Patrick. *Texaco*. New York: Vintage Books, 1998.

Chavez, Daniel. "Sweat Equity: How Uruguay's Housing Coops Provide Sol-
    idarity and Shelter to Low-Income Families." P2P Foundation, June 21,
    2018. https://blog.p2pfoundation.net/sweat-equity-how-uruguays-hous-
    ing-coops-provide-solidarity-and-shelter-to-low-income-families/2018/06/21.

Clapp, Jennifer, and Walter G. Moseley, "This Food Crisis Is Different:
    COVID-19 and the Fragility of the Neoliberal Food Order." *Journal of
    Peasant Studies* 47, no. 7 (2020): 1393–417.

Cloete, Karl. "Op-Ed: Numsa Supports a Transition from Dirty Energy to
    Clean Renewable Energy." *Daily Maverick*, March 15, 2018. https://www.
    dailymaverick.co.za/article/2018-03-15-op-ed-numsa-supports-a-transi-
    tion-from-dirty-energy-to-clean-renewable-energy/.

Coalition of Climate Justice Movements. "COP25, Social Movements and
    Climate Justice." *The Ecologist*, December 2, 2019. https://theecologist.
    org/2019/dec/02/cop25-social-movements-and-climate-justice.

Cohen, Lizabeth. *A Consumers' Republic: The Politics of Mass Consumption in
    Postwar America*. New York: Knopf, 2007.

Collado, José and Han-Hsiang Wang. "Slum Upgrading and Climate Change
    Adaptation and Mitigation: Lessons from Latin America." *Cities* 104 (Sep-
    tember 2020). https://doi.org/10.1016/j.cities.2020.102791.

Cooper, Melinda. *Life as Surplus: Biotechnology and Capitalism in the Neoliberal
    Era*. Seattle: University of Washington Press. 2008.

COP26. "Political Declaration on the Just Energy Transition in South Afri-
    ca." UN Climate Change Conference (COP26) at the SEC, November 2,
    2021. https://ukcop26.org/political-declaration-on-the-just-energy-transi-
    tion-in-south-africa/.

COSATU Central Executive Committee. "Cosatu: Congress of South Afri-
    can Trade Unions Policy Framework on Climate Change (19/11/2011)."
    November 19, 2011.

Counsell, Simon. "Anatomy of a 'Nature-Based Solution': Total Oil, 40,000 Hectares of Disappearing Savannah, Emmanuel Macron, Norwegian and French 'Aid' to an Election-Rigging Dictator, Trees to Burn, Secret Contracts, and Dumbstruck Conservationists." *REDD Monitor*, April 6, 2021. https://redd-monitor.org/2021/04/16/anatomy-of-a-nature-based-solution-total-oil-40000-hectares-of-disappearing-african-savannah-emmanuel-macron-norwegian-and-french-aid-to-an-election-rigging-dictator-trees/.

Cressey, Daniel. "Widely Used Herbicide Linked to Cancer." *Nature News*, March 24, 2015. https://doi.org/10.1038/nature.2015.17181.

Dale, Gareth, Manu V. Mathai, and Jose A. Puppim de Olivera, eds. *Green Growth: Ideology, Political Economy and the Alternatives*. London: Zed Books, 2016.

Davis, Mike. *Late Victorian Holocausts: El Niño Famines and the Making of the Third World*. New York: Verso Books, 2017.

———. *Planet of Slums*. New York: Verso Books, 2006.

———. "Who Will Build the Ark?" *New Left Review* 61 (January 2010): 29–46. https://newleftreview.org/issues/ii61/articles/mike-davis-who-will-build-the-ark.

Davis, Steven J., et al. "Emissions Rebound from the COVID-19 Pandemic." *Nature Climate Change* 12, no. 5 (May 2022): 412–14. https://doi.org/10.1038/s41558-022-01332-6.

Dawson, Ashley. "Action Strategies Working Group." From "The People's Conference on Climate Change." *Periscope, Social Text Online*, July 12, 2011. https://socialtextjournal.org/periscope_article/action_strategies_working_group_part_1/.

———. "Cape Town Has a New Apartheid." *Washington Post*, July 10, 2018.

———. "Climate Justice: The Emerging Movement against Green Capitalism." *South Atlantic Quarterly* 109, no. 2 (April 2010): 313–38. https://doi.org/10.1215/00382876-2009-036.

———. *Extinction: A Radical History*. New York: OR Books, 2016.

———. *Extreme Cities: The Peril and Promise of Urban Life in the Age of Climate Change*. New York: Verso Books, 2019.

———. *People's Power: Reclaiming the Energy Commons*. New York: OR Books, 2020.

———. "Why We Need a Global Green New Deal." *New Politics* 12, no. 4 (2010).

Dawson, Ashley, and Oscar Olivera. "The Cochabamba Water Wars: An Interview with Oscar Olivera." *Social Text Online*, July 5, 2011. https://socialtextjournal.org/periscope_article/the_cochabamba_water_wars/.

Dednam, Charles. "COVID-19: The South African Steel Industry." Trade and Industrial Policy Strategies, July 2020.

Deleuze, Gilles, and Félix Guattari, *A Thousand Plateaus*. Minneapolis: University of Minnesota Press, 1987.

Desmarais, Annette Aurélie. *La Vía Campesina: Globalization and the Power of Peasants*. Ann Arbor, MI: Pluto, 2007.

Desmond, Matthew. *Evicted: Poverty and Profit in the American City*. New York: Crown Publishers, 2016.

Dhanjal, Swaraj Singh. "Adani Group Raises $9 Billion from Offshore Bond Market." *mint*, October 6, 2021. https://www.livemint.com/companies/news/adani-group-raises-9-billion-from-offshore-bond-market-11633457139162.html.

Dhillon, Jaskiran, and Nick Estes. *Standing with Standing Rock: Voices from the #NoDALPL Movement*. Minneapolis: University of Minnesota Press, 2019.

Dillon, Tom. "Glasgow Deal to Tackle Emissions Includes Nature-Based Solutions." Pew Charitable Trusts, November 18, 2021. https://www.pewtrusts.org/en/research-and-analysis/articles/2021/11/18/glasgow-deal-to-tackle-emissions-includes-nature-based-solutions.

Dinerstein, Eric, et al. "An Ecoregion-Based Approach to Protecting Half the Terrestrial Realm." *Bioscience* 67, no. 6 (June 2017): 534–45.

Dirzo, Rudolfo, et al. "Defaunation in the Anthropocene." *Science* 345, no. 6195 (2014): 401–6. doi: 10.1126/science.1251817.

Doshi, Sapana. "Greening Displacements, Displacing Green: Environmental Subjectivity, Slum Clearance, and the Embodied Political Ecologies of Dispossession in Mumbai." *International Journal of Urban and Regional Research* 42, no. 1 (January 2019): 112–32. https://doi.org/10.1111/1468-2427.12699.

Douwe van der Ploeg, Jan. "Growing Back Stronger: Choosing Resilient Food Systems in the Wake of Covid-19." Transnational Institute, September 17, 2020. https://www.tni.org/en/foodsystems.

Dowie, Mark. *Conservation Refugees: The Hundred-Year Conflict between Global Conservation and Native Peoples*. Cambridge, MA: MIT Press, 2009.

———. "Conservation Refugees." *Orion Magazine*, February 21, 2015. https://orionmagazine.org/article/conservation-refugees/.

Dupont, Veronique. "US 'Superweeds' Epidemic Shines Spotlight on GMOs." *Phys*, January 13, 2014. https://phys.org/news/2014-01-superweeds-epidemic-spotlight-gmos.html.

Edelman, Marc, et al., eds. *Global Land Grabbing and Political Reactions "from Below."* New York: Routledge, 2018.

EJOLT. "Coal Mining Conflict in Hazaribagh with NTPC in Jharkhand, India." Environmental Justice Atlas, October 9, 2016. https://ejatlas.org/conflict/illegal-land-acquisition-for-coal-mining-and-violent-protest-in-hazaribagh-jharkhand.

Ellis-Petersen, Hannah. "India Criticised over Coal at Cop26—but Real Villain Was Climate Injustice." *The Guardian*, November 14, 2021.

Emmanouil, Nia, and Carla Chan Unger. *First Peoples and Land Justice Issues in Australia: Addressing Deficits in Corporate Accountability*. RMIT University, March 2021.

Eskelinen, Teppo, and Rikard Warlenius. "Possibilities and Limits of Green Keynesianism." In *The Politics of Ecosocialism: Transforming Welfare*, edited by Borgnäs Kajsa and Johanna Perkio, 101–15. Oxford: Routledge, 2015.

Eskom Research Reference Group. *Eksom Transformed: Achieving a Just Energy Transition for South Africa*, July 2020.

Estes, Nick. *Our History Is the Future: Standing Rock versus the Dakota Access Pipeline, and the Long Tradition of Indigenous Resistance*. London: Verso Books, 2021.

Fabricant, Nicole, and Bret Gustafson. "Revolutionary Extractivism in Bolivia?" *NACLA*, March 2, 2015. https://nacla.org/news/2015/03/02/revolutionary-extractivism-bolivia.

Fanon, Frantz. *The Wretched of the Earth*. New York: Grove Press, 1968.

Featherstone, David. *Solidarity: Hidden Histories and Geographies of Internationalism*. London: Zed Books, 2012.

Ferris, Nick. "Why Critical Mineral Supplies Won't Scupper the Energy Transition." *Energy Monitor*, June 23, 2021. https://www.energymonitor.ai/finance/risk-management/why-we-need-a-level-headed-approach-to-energy-transition-minerals.

Fine, Ben, and Zavareh Rustomjee. *Political Economy of South Africa: From Minerals-Energy Complex to Industrialisation*. New York: Routledge, 1996.

Fisher, Mark. *Capitalist Realism: Is There No Alternative?* Winchester, UK: Zero Books, 2009.

Food and Agriculture Organization. "1.02 Billion People Hungry." Food and Agriculture Organization, June 19, 2009. https://www.fao.org/news/story/pt/item/20568/icode/

———. "Gender: Key to Sustainability and Food Security." http://www.fao.org/News/1997/introG-e.htm.

———. "Impacts of Covid-19 on Food Security and Nutrition: Developing Effective Policy Responses to Address the Hunger and Malnutrition Epidemic," September 2020. https://www.fao.org/3/cb1000en/cb1000en.pdf.

———. "Rome Declaration on World Food Security," November 13–17, 1996.

Friends of the Earth. "Civil Society Groups Release 'Fair Shares NDC' Model." April 8, 2021. https://foe.org/news/civil-society-groups-release-fair-shares-ndc-model-for-revised-u-s-climate-action-pledge-under-the-paris-agreement/#:~:text=The%20U.S.%20Fair%20Shares%20NDC,GHG%20 emissions%20annually%20by%202030.

———. *Junk Agroecology: The Corporate Capture of Agroecology for a Partial Ecological Transition without Social Justice*. Crocevia: The Transnational Institute. 2020.

Fuentes-Bracamontes, Rolando. "Is Unbundling Electricity Services the Way Forward for the Power Sector?" *Electricity Journal* 29, no. 9 (November 2016): 16–20. https://doi.org/10.1016/j.tej.2016.10.006.

Gadgil, Madhav, and Ramachandra Guha. *This Fissured Land: An Ecological History of India*. New York: Oxford University Press, 2012.

Gago, Verónica. *Feminist International: How to Change Everything*. New York: Verso Books, 2020.

Galvao, Louisa Abbott. "Hidden Financing of Fossil Fuels: World Bank and IMF Edition." Friends of the Earth, May 25, 2021. https://foe.org/blog/the-hidden-flows-of-finance-to-fossil-fuels-world-bank-and-imf-edition/.

Gamble, Julie. "A Transit Manifesto for Quito." *NACLA Report on the Americas* 52, no. 2 (June 2020): 199–205. https://doi.org/10.1080/10714839.2020.17 68744.

Gardiner, Beth. "The Deadly Cost of Dirty Air." *National Geographic* 239, no. 4 (April 2021): 40–62.

Gebrial, Dalia, and Harpreet Kaur Paul. *Perspectives on a Global Green New Deal*. London: Rosa Luxemburg Foundation, 2021.

Gidwani, Vinay. *Capital, Interrupted: Agrarian Development and the Politics of Work in India*. Minneapolis: University of Minnesota Press, 2008.

Giliam, Stefan, et al. "Global Patterns of Material Flows and Their Socio-economic and Environmental Implications: A MFA Study on All Countries World-Wide from 1980 to 2009." *Resources* 3, no. 1 (2014): 319–39.

Gillespie, Tom. "Accumulation by Urban Dispossession: Struggles over Urban Space in Accra, Ghana." *Transactions of the Institute of British Geographers* 41, no.1 (2016): 66–77.

Gleeson, Brendan. *The Urban Condition*. New York: Routledge, 2014.

Global Witness. *Last Line of Defence*. London: Global Witness. 2021.

Goldenberg, Suzanne, and Allegra Stratton. "Barack Obama's Speech Disappoints and Fuels Frustration at Copenhagen." *The Guardian*, December 18, 2009.

Gopalakrishnan, Shankar. "The Forest Rights Act." *Economic and Political Weekly* 52, no. 31 (June 2015): 7–8. https://www.epw.in/journal/2017/31/review-environment-and-development/forest-rights-act.html.

GRAIN. *The Great Climate Robbery: How the Food System Drives Climate Change and What We Can Do about It*. Oxford: New Internationalist Publications, 2016.

Greenfield, Patrick. "Humans Exploiting and Destroying Nature on Unprecedented Scale – Report." *The Guardian*, September 9, 2020.

————. "South African Environmental Activist Shot Dead in Her Home." *The Guardian*, October 23, 2020.

Griswold, Shaun. "Interior Department Report Details the Brutality of Federal Indian Boarding Schools." *Wisconsin Examiner*, May 17, 2022. https://wisconsinexaminer.com/2022/05/17/interior-department-report-details-the-brutality-of-federal-indian-boarding-schools/.

groundWork. *Coal Kills: Research and Dialogue for a Just Transition*. Pietermaritzburg, SA: groundWork, 2018.

Grove, Richard. *Green Imperialism: Colonial Expansion, Tropical Island Edens, and the Origins of Environmentalism, 1600–1860*. New York: Cambridge University Press, 1996.

————. "The Culture of Islands and the History of Environmental Concern." In *Climate Change and the Humanities: Historical, Philosophical and Interdisciplinary Approaches to the Contemporary Environmental Crisis*, edited by Alexander Elliott, James Cullis, and Vinita Damodaran, 69–92. New York: Palgrave Macmillan, 2017.

Grunwald, Michael. *The New New Deal: The Hidden Story of Change in the Obama Era*. New York: Simon & Schuster Paperbacks, 2013.

Grzincic, Barbara. "Indian Coal Plant's World Bank Lender Immune from Enviro Suit." *Reuters*, July 7, 2021. https://www.reuters.com/legal/transactional/indian-coal-plants-world-bank-lender-immune-enviro-suit-2021-07-07/.

Guha, Ramachandra, and Juan Martínez Alier. *Varieties of Environmentalism: Essays North and South*. Earthscan, 1997.

Guha, Ranajit. *A Rule of Property for Bengal: An Essay on the Idea of Permanent Settlement*. Durham, NC: Duke University Press, 1996.

Harvey, Chelsea. "Climate Pledges Still Not Enough to Keep Warming below 2-Degree Limit." *Scientific American*, November 23, 2021. https://www.scientificamerican.com/article/climate-pledges-still-not-enough-to-keep-warming-below-2-degree-limit/.

Harvey, David. *A Brief History of Neoliberalism*. New York: Oxford University Press, 2005.

————. *The New Imperialism*. New York: Oxford University Press, 2003.

————. "Why Marx's Capital Still Matters." *Jacobin*, July 12, 2018.

Heer, Jeet. "After the El Paso Massacre, the Choice Is Green Socialism or Eco-Fascism." *The Nation*, August 7, 2019. https://www.thenation.com/article/archive/el-paso-mass-shooting-fascism/.

Henze, Veronika. "Global Investment in Low-Carbon Energy Transition Hit $755 Billion in 2021." *BloombergNEF*, January 27, 2022. https://about.bnef.com/blog/global-investment-in-low-carbon-energy-transition-hit-755-billion-in-2021/.

Hickel, Jason. "Extreme Poverty Isn't Natural, It's Created." *Jason Hickel* (blog), March 28, 2021. https://www.jasonhickel.org/blog.

———. "The Contradiction of the Sustainable Development Goals: Growth versus Ecology on a Finite Planet." *Sustainable Development* 27, no. 5 (September/October 2019): 873–84.

Hill, David. "Ecuador Pursued China Oil Deal While Pledging to Protect Yasuni, Papers Show." *The Guardian*, February 19, 2014,

"How a Just Transition Can Make India's Coal History." BBC, November 9, 2021. https://www.bbc.com/future/article/20211103-india-how-a-just-transition-can-make-coal-history.

Holston, James. *Insurgent Citizenship: Disjunctions of Democracy and Modernity in Brazil*. Princeton, NJ: Princeton University Press, 2007.

Holt-Giménez, Eric. *A Foodie's Guide to Capitalism: Understanding the Political Economy of What We Eat*. New York: NYU Press, 2017.

Hot City Collective. "Hot City: Compound Crisis and Popular Struggle in NYC." *Verso Books* (blog), August 3, 2020. https://www.versobooks.com/blogs/4811-hot-city-compound-crisis-and-popular-struggle-in-nyc.

Imbs, Jean. "The Premature Deindustrialization of South Africa." In *The Industrial Policy Revolution II: Africa in the Twenty-first Century*, edited by Joseph E. Stiglitz, Justin Yifu Lin, and Ebrahim Patel. International Economic Association Series. London: Palgrave Macmillan, 2013. https://doi.org/10.1057/9781137335234_20.

Im, Eun-Soon, Jeremy Pal, and Elfatih Eltahir. "Deadly Heat Waves Projected in the Densely Populated Agricultural Regions of South Asia." *Science Advances* 3, no. 8 (August 2017). http://doi.org/10.1126/sciadv.1603322.

Immerwahr, Daniel, "The Politics of Architecture and Urbanism in Postcolonial Lagos, 1960–1986." *Journal of African Cultural Studies* 19, no. 2 (December 2007): 165-86.

Imran, Zafar. "Climate Change in the Indian Farmers' Protest." *Le Monde diplomatique*, February 1, 2021. https://mondediplo.com/outsidein/climate-indian-farmers.

Indigenous Environmental Network. "Carbon Offsets Cause Conflict and Colonialism." May 18, 2016.

———. "Exposing REDD: The False Climate Solution." October 22, 2012.

———. "Just Transition." https://www.ienearth.org/justtransition/.

Intergovernmental Science-Policy Platform on Biodiversity and Ecosystem Services. "Media Release: Nature's Dangerous Decline 'Unprecedented'; Species Extinction Rates 'Accelerating.'" Press Release. May 7, 2019. https://ipbes.net/news/Media-Release-Global-Assessment.

International Energy Association. "India Has the Opportunity to Build a New

Energy Future." February 9, 2021. https://www.iea.org/news/india-has-the-opportunity-to-build-a-new-energy-future.

International Land Coalition. *Tirana Declaration*. Global Assembly 2011.

International Renewable Energy Association. *RE-organising Power Systems for the Transition*. IRENA, June 2022. https://www.irena.org/publications/2022/Jun/RE-organising-Power-Systems-for-the-Transition

———. *Tracking SDG 7: The Energy Progress Report 2022*. Washington, DC: World Bank, 2022.

International Union for Conservation of Nature. *The Marseille Manifesto*, September 2021. https://www.iucncongress2020.org/programme/marseille-manifesto.

———. "Nature-based Solutions." (Website) https://www.iucn.org/theme/nature-based-solutions/about.

Jones, Van. "Working Together for a Green New Deal." *The Nation*, October 28, 2008. https://www.thenation.com/article/archive/working-together-green-new-deal/.

Just Transition Initiative Team. "Understanding Just Transitions in Coal Dependent Communities." October 2021. https://justtransitioninitiative.org/understanding-just-transitions-in-coal-dependent-communities/.

Kaika, Maria. *City of Flows: Modernity, Nature, and the City*. New York: Routledge, 2004.

Kamanzi, Brian. "Collapse of Energy Utilities Sounds a Warning Bell." *New Frame*, February 18, 2022. https://www.newframe.com/collapse-of-energy-utilities-sounds-a-warning-bell/.

Kaziranga National Park. "History of Kaziranga National Park." https://www.kaziranga-national-park.com/kaziranga-history.shtml.

King, Robin, et al. *Confronting the Urban Housing Crisis in the Global South: Adequate, Secure, and Affordable Housing*. World Resources Institute, July 2017. https://www.wri.org/research/confronting-urban-housing-crisis-global-al-south-adequate-secure-and-affordable-housing.

Klein, Naomi. "Let Them Drown." *London Review of Books*, November 1, 2021. https://www.lrb.co.uk/the-paper/v38/n11/naomi-klein/let-them-drown.

———. *This Changes Everything: Capitalism versus the Climate*. New York: Simon & Schuster, 2014.

Kolbert, Elizabeth. *The Sixth Extinction: An Unnatural History*. New York: Henry Holt, 2014.

Kotz, David. "End of the Neoliberal Era? Crisis and Restructuring in American Capitalism." *New Left Review* 113 (2018): 29–55.

Koven, Joel. *The Enemy of Nature: The End of Capitalism or the End of the World*. London: Zed, 2007.

Krausmann, Fridolin, et al. "Growth in Global Materials Use, GDP and Population during the 20th Century." *Ecological Economics* 68, no. 10 (2009): 2696–705.

Krugman, Paul. "Did Democrats Just Save Civilization?" *New York Times*, August 8, 2022. https://www.nytimes.com/2022/08/08/opinion/climate-inflation-bill.html.

Kunstler, James Howard. *The Long Emergency: Surviving the Converging Crises of the Twenty-First Century*. New York: Grove/Atlantic, 2005.

Lakhani, Nina. "Landmark US Climate Bill Will Do More Harm than Good, Groups Say." *The Guardian*, August 9, 2022.

Lavelle, Marianne. "2016: Obama's Climate Legacy Marked by Triumphs and Lost Opportunities." *Inside Climate News*, December 7, 2020. https://insideclimatenews.org/news/26122016/obama-climate-change-legacy-trump-policies/.

La Via Campesina. "Food Sovereignty." January 15, 2003. https://viacampesina.org/en/what-are-we-fighting-for/food-sovereignty-and-trade/.

———. "Globalize Hope: New Film on the History of La Via Campesina." 2020. https://tv.viacampesina.org/Globalize-Hope?lang=en.

———. "International Conference of Agrarian Reform: Marabá Declaration." April 22, 2016. https://viacampesina.org/en/international-conference-of-agrarian-reform-declaration-of-maraba1/.

———. "Opinion: Agroecology for Gender Equality." September 20, 2016. https://viacampesina.org/en/opinion-agroecology-for-gender-equality/.

———. "Sow the Seeds of Struggle and Resistance, and Cultivate our Rights!" July 21, 2020. https://viacampesina.org/en/la-via-campesina-says-its-time-totransform/.

———. "Struggles of La Via Campesina | For Agrarian Reform and the Defense of Life, Land and Territories." October 16, 2017. https://viacampesina.org/en/struggles-la-via-campesina-agrarian-reform-defense-life-land-territories/.

Lawlor, Mary. *Final Warning: Death Threats and Killings of Human Rights Defenders: Report of the Special Rapporteur on the Situation of Human Rights Defenders*. Geneva, Switzerland: UN, 2020.

Lawrence, Mathew, and Laurie Laybourn-Langton. *Planet on Fire: A Manifesto for the Age of Environmental Breakdown*. New York: Verso Books, 2022.

Leber, Rebecca. "The US Finally Has a Law to Tackle Climate Change." *Vox*, August 16, 2022. https://www.vox.com/policy-and-politics/2022/7/28/23281757/whats-in-climate-bill-inflation-reduction-act.

Lenin, Vladimir. *Imperialism: The Highest Stage of Capitalism*. Reprint, New York: Penguin Classics, 2010 [1917].

Liberti, Stefano. *Land Grabbing: Journeys in the New Colonialism.* New York: Verso Books, 2013.

Lima, Márcia Maria Tait, and Vanessa Brito de Jesus. "Questions about Gender and Technology in Agroecology." *Scientiae Studia* 15, no. 1 (November 2018): 73–96. https://doi.org/10.11606/51678-31662017000100005.

Linebaugh, Peter. "Karl Marx, the Theft of Wood and Working-Class Composition: A Contribution to the Current Debate." *Crime and Social Justice* 6 (Fall–Winter 1979): 1–29.

Locke, John. *Second Treatise of Government.* Edited by Thomas P. Peardon. New York: Liberal Arts Press, 1952.

Longo, Fiore. "Why Nature-Based Solutions Won't Solve the Climate Crisis— They'll Just Make Rich People Even Richer." *Common Dreams*, October 13, 2021. https://www.commondreams.org/views/2021/10/13/why-nature-based-solutions-wont-solve-climate-crisis-theyll-just-make-rich-people.

Lorde, Audre. "The Master's Tools Will Never Dismantle the Master's House." In *Sister Outsider: Essays and Speeches*, 110–14. Berkeley, CA: Crossing Press, 1984.

Luxemburg, Rosa. *The Junius Pamphlet.* 1916. https://www.marxists.org/archive/luxemburg/1915/junius/.

Maclean, Ruth, and Dionne Searcey. "Congo to Auction Land to Oil Companies: 'Our Priority Is Not to Save the Planet.'" *New York Times*, July 24, 2022.

Madeley, John. *Hungry for Trade.* London: Zed. 2000.

Maggott, Terri, et al. *Energy Racism: The Electricity Crisis and the Working Class in South Africa.* Centre for Sociological Research and Practice. April 2022.

Malm, Andreas. *Corona, Climate, Chronic Emergency: War Communism in the Twenty-First Century.* London: Verso Books, 2020.

———. "Tahrir Submerged? Five Theses on Revolution in the Age of Climate Change." *Capitalism Nature Socialism* 25, no. 3 (March 2014): 28–44. https://doi.org/10.1080/10455752.2014.891629.

———. *The Progress of This Storm: Nature and Society in a Warming World.* New York: Verso Books, 2017.

———. *White Skin, Black Fuel: On the Danger of Fossil Fascism.* London: Verso Books, 2021.

Mander, Harsh. "'Urban Maoists': A Curious New Creature in Modi's India." *South China Morning Post*, August 31, 2018. https://www.scmp.com/week-asia/politics/article/2162232/urban-maoists-modis-india-if-you-are-right-you-must-be-left.

Marais, Hein. *South Africa Pushed to the Limit: The Political Economy of Change.* London: Zed Books, 2013.

Marx, Karl. *Capital.* Vol. 1, translated by Ben Fowkes. London: Vintage, 1976 [1887].

Mashal, Mujib, Emily Schmall, and Hari Kumar. "As Angry Farmers Take to New Delhi's Streets, Protests Turn Violent." *New York Times,* January 25, 2021.

Mason, Paul. *Postcapitalism: A Guide to Our Future.* New York: Farrar, Straus and Giroux, 2015.

Masterson, Victoria. "'Renewables' Power Ahead to Become the World's Cheapest Source of Energy in 2020." World Economic Forum, July 5, 2021. https://www.weforum.org/agenda/2021/07/renewables-cheapest-energy-source/.

Mathiesen, Karl. "The Last-Minute Coal Demand That Almost Sunk the Glasgow Climate Deal." *Politico,* November 13, 2021. https://www.politico.com/news/2021/11/13/coalglasgow-climate-deal-521802.

Mayer, Andreas, and Haas, Willi. "Cumulative Material Flows Provide Indicators to Quantify the Ecological Debt." *Journal of Political Ecology* 23 (2016): 350–63.

Mayersohn, Norman. "How High Tech is Transforming One of the Oldest Jobs: Farming." *New York Times,* September 6, 2019.

McAdam, Jane. *Climate Change, Forced Migration, and International Law.* New York: Oxford University Press, 2012.

McFarlane, Collin. "The Entrepreneurial Slum: Civil Society, Mobility and the Co-production of Urban Development." *Urban Studies* 49, no. 13 (September 2012): 2795–816. https://doi.org/10.1177/0042098012452460.

McGregor, Deborah, Steven Whitaker, and Mahisha Sritharan. "Indigenous Environmental Justice and Sustainability." *Current Opinion in Environmental Sustainability* 43 (2020): 35–40. https://doi.org/10.1016/j.cosust.2020.01.007.

McGuirk, Justin. *Radical Cities: Across Latin America in Search of a New Architecture.* New York: Verso Books, 2014.

McKibben, Bill. "Bad News for Obama: Fracking May Be Worse Than Burning Coal." *Mother Jones,* September 8, 2014. https://www.motherjones.com/environment/2014/09/methane-fracking-obama-climate-change-bill-mckibben/.

Mehrotra, Karishma. "In India's Coal Belts, Jobs Are Now Hard to Get—and Harder to Keep." *Scroll.in,* February 10, 2022. https://scroll.in/article/1016936/in-indias-coal-belts-jobs-are-now-hard-to-get-and-harder-to-keep.

Merchant, Carolyn. *The Death of Nature: Women, Ecology, and the Scientific Revolution.* New York: HarperOne, 1990.

Mignolo, Walter D., and Catherine Walsh. *On Decoloniality: Concepts, Analytics, Praxis.* Durham, NC: Duke University Press, 2018.

Milman, Oliver. "A Closer Look at Joe Manchin's Ties to the Fossil Fuel Industry." *Mother Jones*, October 21, 2021. https://www.motherjones.com/politics/2021/10/a-closer-look-at-joe-manchins-ties-to-the-fossil-fuel-industy/.

Mintz, Sidney. *Sweetness and Power: The Place of Sugar in Modern History*. New York: Penguin, 1986.

Mitchell, Timothy. *Carbon Democracy: Political Power in the Age of Oil*. London: Verso Books, 2013.

Mitman, Gregg, Donna Haraway, and Anna Tsing. "Reflections on the Plantationocene: A Conversation with Donna Haraway and Anna Tsing." *Edge Effects Magazine*, October 12, 2019.

Mohanty, Abinash. "Preparing India for Extreme Climate Events." CEEW, December 2020. https://www.ceew.in/publications/preparing-india-for-extreme-climate-weather-events.

Mohanty, Abinash, and Shreya Wadhawan. "Mapping India's Climate Vulnerability: A District-Level Assessment." *Hindustan Times*, November 11, 2021. https://www.hindustantimes.com/ht-insight/climate-change/mapping-india-s-climate-vulnerability-a-district-level-assessment-101636642145178.html.

Monbiot, George. "How Labour Could Lead the Global Economy Out of the 20th Century." *The Guardian*, October 11, 2017.

———. *Out of the Wreckage: A New Politics for an Age of Crisis*. New York: Verso Books, 2017.

Montgomery, David. "Peak Soil." *New Internationalist*, December 1, 2008. https://newint.org/features/2008/12/01/soil-depletion.

Montgomery, David R. *Dirt: The Erosion of Civilization*. Berkeley, CA: University of California Press, 2007.

———. *Growing a Revolution: Bringing Our Soil Back to Life*. New York: W. W. Norton, 2017.

Moore, James W, ed. *Anthropocene or Capitalocene?: Nature, History, and the Crisis of Capitalism*. Oakland, CA: PM Press/Kairos. 2016.

Mora, Camilo, et al. "Global Risk of Deadly Heat." *Nature Climate Change* 7 (2017): 501–6. https://doi.org/10.1038/nclimate3322.

Morgan, Jamie. "The Fourth Industrial Revolution Could Lead to a Dark Future." *The Conversation*, January 9, 2020.

Morris, Craig, and Arne Jungjohann. *Energy Democracy: Germany's Energiewende to Renewables*. New York: Springer International, 2018.

Moseley, William G., and Jane Battersby. "The Vulnerability and Resilience of African Food Systems, Food Security and Nutrition in the Context of the COVID-19 Pandemic." *African Studies Review* 63, no. 3 (2020): 1–13.

Mottiar, Shauna. "Shaping a Township: Self-Connecting as Counter-Conduct in Umlazi, Durban." *Journal of the British Academy* 9, no. 11 (2021): 93–106.

Muchhala, Bhumika. "Towards a Decolonial and Feminist Global Green New Deal." Rosa-Luxemburg-Stiftung, August 24, 2020. https://www.rosalux. de/en/news/id/43146/towards-a-decolonial-and-feminist-global-green-new-deal.

Mukpo, Ashoka. "In the DRC's Forests, a Tug-of-War between Oil and Aid." *Mongabay*, June 7, 2022. https://news.mongabay.com/2022/06/in-the-drcs-forests-a-tug-of-war-between-oil-and-aid/.

Murphy, Caryle. "To Cope with Embargoes, S. Africa Converts Coal into Oil." *Washington Post*, April 27, 1979. https://www.washingtonpost.com/archive/politics/1979/04/27/to-cope-with-embargoes-s-africa-converts-coal-into-oil/cd39adab-5084-4e46-a28f-79de2896f75e/.

Muttitt, Greg, and Sivan Kartha. "Equity, Climate Justice and Fossil Fuel Extraction: Principles for a Managed Phase Out." *Climate Policy* 20, no. 8 (September 2020): 1024–42. https://doi.org/10.1080/14693062.2020.1763900.

Muyskens, John, and Juliet Eilperin. "Biden Calls for 100 Percent Clean Electricity by 2035. Here's How Far We Have to Go." *Washington Post*, July 30, 2020.

Naidoo, Prinesha, and Felix Njini. "Iconic South African Mines Ravaged Economy's Unlikely Savior." *Bloomberg*, July 6, 2021. https://www.bloomberg.com/news/articles/2021-07-06/iconic-south-african-mines-are-ravaged-economy-s-unlikely-savior.

National Sample Survey Office. *India: Common Property Resources, Sanitation, and Hygiene Services, NSS 54th Round: January–June 1998*. Government of India, March 16, 2016. mospi.nic.in/rept%20_%20pubn/452_final.pdf

Navdanya International. *Ag One: The Recolonization of Agriculture* (Navdanya/RESTE, 2020).

Negri, Antonio. *Marx beyond Marx*. New York: Autonomedia/Pluto, 1991.

NET Web Desk. "Assam: Commandos Will Be Deployed in Kaziranga National Park to Combat Rhino Poaching." *Northeast Today*, January 24, 2022. https://www.northeasttoday.in/2022/01/24/assam-commandos-will-be-deployed-in-kaziranga-national-park-to-combat-rhino-poaching/.

Neumann, Roderick P. "Dukes, Earls, and Ersatz Edens: Aristocratic Nature Preservationists in Colonial Africa." *Environment and Planning D: Society and Space* 14, no. 1 (1996): 79–98.

Neuwirth, Robert. *Shadow Cities: A Billion Squatters, a New Urban World*. New York: Routledge, 2006.

Ngwane, Trevor, Luke Sinwell, and Immanuel Ness. *Urban Revolt: State Power and the Rise of People's Movements in the Global South*. Chicago: Haymarket Books, 2017.

Niekerk, Ashley van, et al. "Energy Impoverishment and Burns: The Case for an Expedited, Safe and Inclusive Energy Transition in South Africa." *South African Journal of Science* 118, no. 3/4 (March 29, 2022). https://doi.org/10.17159/sajs.2022/13148.

Niiler, Eric. "Why Gene Editing Is the Next Food Revolution." *National Geographic*, August 10, 2018.

Nixon, Rob. *Slow Violence and the Environmentalism of the Poor*. Cambridge, MA: Harvard University Press, 2013.

NUMSA. "Motivations for a Socially-Owned Renewable Energy Sector," October 15, 2012. https://numsa.org.za/2012/10/motivations-for-a-socially-owned-renewable-energy-sector-2012-10-15/.

Olivera, Oscar, and Tom Lewis. *¡Cochabamba! Water War in Bolivia*. Cambridge, MA: South End Press, 2004.

Oskarsson, Patrik, et al. "India's New Coal Geography: Coastal Transformations, Imported Fuel and State-Business Collaboration in the Transition to More Fossil Fuel Energy." *Energy Research & Social Science* 73 (March 2021): 1. https://doi.org/10.1016/j.erss.2020.101903.

Out of the Woods Collective. *Hope against Hope: Writings on Ecological Crisis*. Philadelphia: Common Notions, 2020.

Oxfam. "Poorer Nations Expected to Face up to $75 Billion Six-Year Shortfall in Climate Finance." September 19, 2021. https://www.oxfamamerica.org/press/poorer-nations-expected-to-face-up-to-75-billion-six-year-shortfall-in-climate-finance-oxfam/.

Oxfam, International Land Coalition, and Rights and Resources Initiative. *Common Ground: Securing Land Rights and Safeguarding the Earth*. Oxford: Oxfam. March 2016. https://rightsandresources.org/wp-content/uploads/2016/04/Global-Call-to-Action_Common-Ground_Land-Rights_April-2-16_English.pdf.

Pal, Sumedha. "One Year after Arrest, Organisations Demand Adivasi Activist Hidme Markam's Release." *The Wire*, March 9, 2022. https://thewire.in/rights/one-year-after-arrest-organisations-demand-adivasi-activist-hidme-markams-release.

Pardikar, Rishika. "Global North Is Responsible for 92% of Excess Emissions." *EOS*, October 28, 2020. https://eos.org/articles/global-north-is-responsible-for-92-of-excess-emissions.

Parenti, Christian. *Tropics of Chaos: Climate Change and the New Geography of Violence*. New York: Nation Books, 2012.

Paret, Marcel. "Resistance within South Africa's Passive Revolution: From Racial Inclusion to Fractured Militancy." *International Journal of Politics, Culture, and Society*, November 6, 2021. https://doi.org/10.1007/s10767-021-09410-x.

Patel, Raj, and Jim Goodman. "A Green New Deal for Agriculture." *Jacobin*,

April 4, 2019. https://jacobin.com/2019/04/green-new-deal-agriculture-farm-workers.

Patel, Raj, and Philip McMichael. "A Political Economy of the Food Riot." *Review, A Journal of the Fernand Braudel Center* 32, no. 1 (2009): 9–35.

Patel, Rajeev. "International Agrarian Restructuring and the Practical Ethics of Peasant Movement Solidarity." *Journal of African and Asian Studies* 41, no. 1/2 (2006): 71–93.

Pegg, David. "Why the Mundra Power Plant Has Given Tata a Mega Headache." *The Guardian*, April 16, 2015

Pettifor, Ann. *The Case for the Green New Deal*. New York: Verso Books, 2020.

Perfecto, Ivette, John Vandermeer, and Angus Wright. *Nature's Matrix: Linking Agriculture, Biodiversity Conservation, and Food Sovereignty*, 2nd ed. New York: Routledge, 2019.

Pitzer, Andrea. "Concentration Camps Existed Long before Auschwitz." *Smithsonian* (magazine), November 2, 2017. https://www.smithsonianmag.com/history/concentration-camps-existed-long-before-Auschwitz-180967049/.

Podur, Justin. "Leaving Behind the Racist and Imperialist Baggage of the Original New Deal." In *Perspectives on a Global Green New Deal*, edited by Harpreet Kaur Paul and Dalia Gebrial, 123–26. London: Rosa-Luxemburg-Stiftung, 2021.

Powell, B.H. Baden. *Forest Law*. London: Bradbury Agnaw and Co., 1893.

Press Information Bureau Delhi. "Unleashing Coal: New Hopes for Atmanirbhar Bharat." Press Release, June 11, 2020. https://pib.gov.in/PressReleasePage.aspx?PRID=1630919.

Protected Planet. "Executive Summary." *Protected Planet Report 2020*, May 2021. https://livereport.protectedplanet.net/chapter-1.

Puffert, Douglas. "Path Dependence." EH.Net Encyclopedia, February 10, 2008. http://eh.net/encyclopedia/path-dependence/.

Putul, Alok Prakash. "As India Faces Coal Shortages, a Mine Extension Has Been Approved in the Pristine Hasdeo Forests." *Scroll.in*, May 29, 2022. https://scroll.in/article/1024834/as-india-faces-coal-shortages-a-mine-extension-has-been-approved-in-the-pristine-hasdeo-forests.

Rainforest Foundation UK. "The 'Post-2020 Global Biodiversity Framework' – How the CBD Drive to Protect 30 Percent of the Planet Could Dispossess Millions." Mapping for Rights, July 2020. https://www.mappingforrights.org/MFR-resources/mapstory/cbddrive/300_million_at_risk_from_cbd_drive.

Rall, Katharina, and Ramin Pejan. *"We Know Our Lives Are in Danger": Environment of Fear in South Africa's Mining-Affected Communities*. New York: Human Rights Watch, 2019.

Raymond, Colin, Tom Matthews, and Radley M. Horton. "The Emergence of Heat and Humidity Too Severe for Human Tolerance." *Science Advances* 6, no. 19 (May 2020). http://www.doi.org/10.1126/sciadv.aaw1838.

The Red Nation. *The Red Deal: Indigenous Action to Save Our Earth*. Philadelphia: Common Notions, 2021.

Reed, Drew. "How Curitiba's BRT Stations Sparked a Transport Revolution—a History of Cities in 50 Buildings, Day 43." *The Guardian*, May 26, 2015.

Rehman, Asad. "Opinion: The 'Green New Deal' Supported by Ocasio-Cortez and Corbyn Is Just a New Form of Colonialism." *The Independent*, May 4, 2019.

Rensberger, Boyce. "Experts Ask Action to Avoid Millions of Deaths in Food Crisis." *New York Times*, July 26, 1974.

Reuters. "UN Chief Urges Wealthy Nations to Phase Out Coal Use by 2030." *Mining Weekly*, March 3, 2021. https://www.miningweekly.com/article/un-chief-urges-wealthy-nations-to-phase-out-coal-use-by-2030-2021-03-03/rep_id:3650.

Roberts, David. "The Key to Tackling Climate Change: Electrify Everything." *Vox*, September 19, 2016. https://www.vox.com/2016/9/19/12938086/electrify-everything.

Robinson, Jennifer. *Ordinary Cities between Modernity and Development*. London: Taylor & Francis, 2006.

Robinson, Kim Stanley. *The Ministry for the Future*. London: Orbit, 2020.

Rosset, Peter M., and Miguel A. Altieri. *Agroecology: Science and Politics*. Halifax, Nova Scotia: Fernwood Publishing, 2017.

Roussinos, Aris. "The Age of Empire Is Back." *UnHerd*, February 17, 2021. https://unherd.com/2021/02/the-rise-of-green-imperialism/.

Rowlatt, Justin. "Kaziranga: The Park That Shoots People to Protect Rhinos." *BBC News*, February 10, 2017. https://www.bbc.com/news/world-south-asia-38909512.

Royal Botanical Gardens, Kew. "State of the World's Plants and Fungi 2020." https://www.kew.org/science/state-of-the-worlds-plants-and-fungi.

Roy, Brototi, and Joan Martinez-Alier. "Environmental Justice Movements in India: An Analysis of the Multiple Manifestations of Violence." *Ecology, Economy and Society – the INSEE Journal* 2, no. 1 (January 2019): 77–92. https://doi.org/10.37773/ees.v2i1.56.

Roy, Brototi, and Anke Schaffartzik. "Talk Renewables, Walk Coal: The Paradox of India's Energy Transition." *Ecological Economics* 180 (February 2021). https://doi.org/10.1016/j.ecolecon.2020.106871.

Ruffini, Antonio. "The Decline of South African Gold Mining." *Engineering and Mining Journal*, June 2010. https://www.e-mj.com/features/the-decline-of-south-african-gold-mining/.

Ryan, Frances. "The Tories Never Cared about Eliminating the Deficit. It Was Just a Pretext to Slash the State." *The Guardian*, March 8, 2018.

Said, Edward W. *Orientalism*. New York: Vintage, 1979.

Saikia, Arunabh. "Kaziranga Activists Jailed: Colleagues Claim This Is Vendetta for Their Role in BBC Film on Poaching." *Scroll.in*, May 5, 2017. https://scroll.in/article/836565/kaziranga-activists-jailed-were-they-held-for-featuring-in-bbc-film-critical-of-poaching-policy.

Salari, Mahmoud, Roxana J. Javid, and Hamid Noghanibehambari. "The Nexus Between $CO_2$ Emissions, Energy Consumption, and Economic Growth in the US." *Economic Analysis and Policy* 69 (March 2021): 182–94.

Sammon, Alexander. "How the Bank Bailout Hobbled the Climate Fight." *New Republic*, October 22, 2018. https://newrepublic.com/article/151700/bank-bailout-hobbled-climate-fight.

Sassen, Saskia. *Expulsions: Brutality and Complexity in the Global Economy*. Cambridge, MA: Belknap Press, 2014.

Satgar, Vishwas. "Reclaiming the South African Dream." *Socialist Project*, January 2, 2012. https://socialistproject.ca/2012/01/b584/.

Save Our Wilderness. "Zac's Christmas Contamination Crisis." January 4, 2022. https://saveourwilderness.org/2021/12/31/zacs-christmas-contamination-crisis/.

Scheidel, Arnim, et al. "Environmental Conflicts and Defenders: A Global Overview." *Global Environmental Change* 63 (July 2020). https://doi.org/10.1016/j.gloenvcha.2020.102104.

Schleicher, Judith, et al. "Protecting Half the Planet Could Directly Affect over One Billion People." *Nature Sustainability* 2 (December 2019): 1094–96.

Schneider, Keith. "World Bank, Despite Promises, Finances Big Coal and Industrial Projects That Threaten Water, Communities." *Circle of Blue* (blog), October 18, 2016. https://www.circleofblue.org/2016/world/world-bank-despite-promises-finances-big-coal-industrial-projects-threaten-water-communities/.

Schwartzman, Peter, and David Schwartzman. *The Earth Is Not for Sale: A Path out of Fossil Capitalism to the Other World That Is Still Possible*. Hackensack, NJ: World Scientific, 2019.

Schwendler, Sônia Fátima, and Lucia Amaranta Thompson. "An Education in Gender and Agroecology in Brazil's Landless Rural Workers' Movement." *Gender and Education* 29, no. 1 (January 2017): 100–114. https://doi.org/10.1080/09540253.2016.1221596.

Scott, James. *Seeing Like a State: How Certain Schemes to Improve the Human Condition Have Failed*. New Haven, CT: Yale University Press, 1999.

Scott, Nick, and Jordan Mendys. "This Is What It Means to Be Poor in India Today." CNN, October 2017. https://www.cnn.com/interactive/2017/10/world/i-on-india-income-gap/.

Sekhri, Abhinav. "How the UAPA Is Perverting the Idea of Justice." *Article14*, July 16, 2022. https://www.article-14.com/post/how-the-uapa-is-perverting-india-s-justice-system.

Shah, Alpa. "The Agrarian Question in a Maoist Guerrilla Zone: Land, Labour and Capital in the Forests and Hills of Jharkhand, India." *Journal of Agrarian Change* 13 (June 2013): 424–50. https://doi.org/10.1111/joac.12027.

Shearing, Cyndie. "Women Count in Agriculture." American Farm Bureau Federation, May 1, 2019. https://www.fb.org/viewpoints/women-count-in-agriculture.

Sheik, Knvul. "A Growing Presence on the Farm: Robots." *New York Times*, February 13, 2020.

Shell PLC. "Nature-Based Solutions." https://www.shell.com/energy-and-innovation/new-energies/nature-based-solutions.html#iframe=L3dlYmF-wcHMvMjAxOV9uYXR1cmVfYmFzZWRfc29sdXRpb25zL3VwZG-F0ZS8.

Sheller, Mimi. *Mobility Justice: The Politics of Movement in an Age of Extremes.* New York: Verso Books, 2018.

Shiva, Vandana. *Who Really Feeds the World.* Berkeley, CA: Zed Books, 2015.

Shivanna, K. R., Rajesh Tandon, and Monika Koul. "'Global Pollinator Crisis' and Its Impact on Crop Productivity and Sustenance of Plant Diversity." In *Reproductive Ecology of Flowering Plants: Patterns and Processes*, edited by Rajesh Tandon, K. R. Shivanna, and Monika Koul, 395–413. Singapore: Springer, 2020. https://doi.org/10.1007/978-981-15-4210-7_16.

Shrivastava, Kumar Sambhav. "Govt to Allow PVT Sector to Manage 40% of Forests." *Hindustan Times*, September 13, 2015. https://www.hindustantimes.com/india/govt-toallow-pvt-sector-to-manage-40-of-forests/story-yOiG4TO4kA2kvykxXNTEBK.html.

Simms, Andrew, et al. *A Green New Deal: Joined-Up Policies to Solve the Triple Crunch of the Credit Crisis, Climate Change and High Oil Prices.* New Economics Foundation, July 2008. https://neweconomics.org/2008/07/green-new-deal.

Simone, AbdouMaliq. "People as Infrastructure: Intersecting Fragments in Johannesburg." *Public Culture* 16, no. 3 (2004): 407–29. muse.jhu.edu/article/173743.

Smadja, Joëlle. "A Chronicle of Law Implementation in Environmental Conflicts: The Case of Kaziranga National Park in Assam (North-East India)."

*South Asia Multidisciplinary Academic Journal* 17 (2018): 1–37.

Smith, Alex, "China's Poorly Planned Cities: Urban Sprawl and the Rural Underclass Left Behind." *SupChina*, March 11, 2020. https://signal.supchina.com/chinas-poorly-planned-cities-urban-sprawl-and-the-rural-underclass-left-behind/.

Smith, Cynthia E. *Design with the Other 90%: Cities*. New York: Cooper-Hewitt, National Design Museum, 2007.

So, Anthony D., et al. "Is Bayh-Dole Good for Developing Countries? Lessons from the US Experience." *PLoS Biology* 6, no. 10 (October 2008). https://doi.org/10.1371/journal.pbio.0060262.

Sontanam, Satya. "All You Want to Know about Coal Mine Auctions." *The Hindu Businessline*, June 22, 2020. https://www.thehindubusinessline.com/opinion/columns/slate/all-you-want-to-know-about/article31892169.ece.

Soper, Kate. *Post-Growth Living: For an Alternative Hedonism*. New York: Verso Books, 2020.

Springer, Jenny, and Fernanda Almeida. *Protected Areas and the Land Rights of Indigenous Peoples and Local Communities*. Rights and Resources Initiative, 2015. https://rightsandresources.org/wp-content/uploads/RRIReport_Protected-Areas-and-Land-Rights_web.pdf.

Srnicek, Nick, and Alex Williams. *Inventing the Future: Postcapitalism and a World without Work*. New York: Verso Books, 2016.

Stahn, Carsten, and Jens Iverson, eds. *Just Peace after Conflict: Jus Post Bellum and the Justice of Peace*. Oxford: Oxford University Press, 2021.

Stebbing, E. P. *The Forests of India*, vol. 2. London: John Lane, 1923.

Stein, Samuel. *Capital City: Gentrification and the Real Estate State*. London: Verso Books, 2019.

Stern, Nicholas. *The Economics of Climate Change: The Stern Review*. New York: Cambridge University Press, 2007.

Stoddard, Ed. "Minerals and Energy: The Decline of South Africa's Mining Sector in Five Charts." *Daily Maverick*, November 23, 2021. https://www.dailymaverick.co.za/article/2021-11-23-the-decline-of-south-africas-mining-sector-in-five-charts/.

StopAdani. "Timeline of #StopAdani Actions." https://www.tiki-toki.com/timeline/entry/1006867/Timeline-of-StopAdani-actions.

Sundar, Nandini. *The Burning Forest: India's War against the Maoists*. London: Verso Books, 2019.

Sweeney, Sean, and John Treat. *The Road Less Travelled: Reclaiming Public Transport for Climate Ready Mobility*. Trade Unions for Energy Democracy, May 2019. https://unionsforenergydemocracy.org/resources/tued-working-papers/tued-working-paper-12/.

Swilling, Mark, et al. "Linking the Energy Transition and Economic Development: A Framework for Analysis of Energy Transitions in the Global South." *Energy Research & Social Science* 90 (August 2022). https://doi.org/10.1016/j.erss.2022.102567.

Táíwò, Olúfẹ́mi O. *Reconsidering Reparations*. New York: Oxford University Press. 2022.

Talukdar, Sushanta. "Waiting for Curzon's Kin to Celebrate Kaziranga." *The Hindu*, January 5, 2005.

Tamburini, Giovanni, et al. "Agricultural Diversification Promotes Multiple Ecosystem Services without Compromising Yield." *Science Advances* 6, no. 45 (November 2020). https://doi.org/10.1126/sciadv.aba1715.

Temper, Leah, et al. "Movements Shaping Climate Futures: A Systematic Mapping of Protests against Fossil Fuel and Low-Carbon Energy Projects." *Environmental Research Letters* 15, no. 12 (2020). https://doi.org/10.1088/1748-9326/abc197.

Transnational Institute, *Rogue Capitalism and the Financialization of Territories and Nature*. September 2020. www.tni.org.

Treuer, David. "Return the National Parks to the Tribes." *The Atlantic*, September 2, 2021. https://www.theatlantic.com/magazine/archive/2021/05/return-the-national-parks-to-the-tribes/618395/.

Tripathi, Bhasker. "India 5th Most Vulnerable to Climate Change Fallouts, Its Poor the Worst Hit." December 5, 2019. https://www.indiaspend.com/india-5th-most-vulnerable-to-climate-change-fallouts-its-poor-the-worst-hit/.

Turner, J. F. C. "The Squatter Settlement: Architecture That Works." *Architectural Design* 38 (1968): 355–60.

UNHCR. "Figures at a Glance." (Webpage) www.unhcr.org/about-unhcr/who-we-are/figures-glance.

United Nations. "A Record 100 Million Forcibly Displaced Worldwide." UN News, May 23, 2022. https://news.un.org/en/story/2022/05/1118772.

———. *The State of Food Security and Nutrition in the World 2020: Transforming Food Systems for Affordable Healthy Diets*. 2020. https://doi.org/10.4060/CA9692EN.

———. "Sustainable Development Goals." (Website) https://sdgs.un.org/goals.

Urban-Think Tank. "Empower Shack." Architizer, December 13, 2020. https://architizer.com/projects/empower-shack/.

US Geological Survey. *Mineral Commodity Summaries 2021*. Reston, VA: US Geological Survey, 2021. https://pubs.usgs.gov/periodicals/mcs2021/mcs2021.pdf.

Vaidyanathan, Rajini. "Climate Change: Why India Can't Live without Coal."

BBC News, September 28, 2021. https://www.bbc.com/news/world-asia-india-58706229.

Vandone, Christopher, and Sheila Khama. "Violence Adds to Uncertainty for South African Mining." Chatham House – International Affairs Think Tank, August 16, 2021. https://www.chathamhouse.org/2021/08/violence-adds-uncertainty-south-africas-mining.

Vasudevan, Alexander. "The Makeshift City: Towards A Global Geography of Squatting." *Progress in Human Geography* 39, no. 3 (April 2014): 338–59.

Venter, Christo, Anjali Mahendra, and Dario Hidalgo. "From Mobility to Access for All: Expanding Urban Transportation Choices in the Global South." World Resources Institute, May 14, 2019. https://www.wri.org/research/mobility-access-all-expanding-urban-transportation-choices-global-south.

Wachsmuth, David, Cohen, Daniel Aldana, and Hillary Angelo. "Expand the Frontiers of Urban Sustainability." *Nature* 536 (2016): 391–93. https://doi.org/10.1038/536391a.

Wallace, Rob. "Dead Epidemiologists: On the Origins of Covid-19." *Monthly Review*, 2020. https://monthlyreview.org/product/dead-epidemiologists-on-the-origins-of-covid-19/.

Wallace-Wells, David. "Hardly Anyone Talks about How Fracking Was an Extraordinary Boondoggle." *New York Times*, July 27, 2022.

Watts, Jonathan. "One Billion People Will Live in Insufferable Heat within 50 Years – Study." *The Guardian*, May 5, 2020.

Weinstein, Liza. "Evictions: Reconceptualizing Housing Insecurity from the Global South." *City & Community* 20, no. 1 (February 2021): 13–23. https://doi.org/10.1111/cico.12503

Welsby, Dan, et al. "Unextractable Fossil Fuels in a 1.5° C World." *Nature* 597, no. 7875 (September 2021): 230–34. https://doi.org/10.1038/s41586-021-03821-8.

Weltz, Adam. "Lumumba Di-Aping: 'We Have Been Asked to Sign a Suicide Pact." *Reimagine: Movements Making Media*, December 14, 2009. https://www.reimaginerpe.org/cj/news/12-9.

Whitehead, Judith. "John Locke, Accumulation by Dispossession and the Governance of Colonial India." *Journal of Contemporary Asia* 42, no. 1 (February 2012): 10–11.

Wijeratna, Alex. *Agroecology: Scaling-Up, Scaling-Out.* ActionAid International, April 2018.

Wilkerson, Jordan. "Why Roundup Ready Crops Have Lost Their Allure." *Science in the News*, October 1, 2017. http://sitn.hms.harvard.edu/flash/2015/roundup-ready-crops/.

Williams, Casey. "Amid Rolling Blackouts, Energy Workers Fight for Clean Public Power in South Africa." *In These Times*, March 31, 2022. https://inthesetimes.com/article/decarbonization-just-transition-labor-unions-south-africa-energy-eskom-workers-privatization.

Williams, Evan Calder. *Combined and Uneven Apocalypse: Luciferian Marxism*. Winchester, UK: Zero Books, 2011.

Williams, Raymond. *The Country and The City*. New York: Oxford University Press, 1973.

Woetzel, Jonathan R. *A Blueprint for Addressing the Global Affordable Housing Challenge*. New York: McKinsey Global Institute, 2014.

WoMin. *No Longer A Life Worth Living*. Johannesburg, SA: WoMin African Gender and Extractives Alliance, 2016.

Woodman, Jo. "In Modi's India, Being a Tribal Woman Is an Act of Resistance." *CounterPunch*, July 9, 2021. https://www.counterpunch.org/2021/07/09/in-modis-india-being-a-tribal-woman-is-an-act-of-resistance/.

World Bank. "CO2 Emissions (Metric Tons per Capita)." https://data.worldbank.org/indicator/EN.ATM.CO2E.PC.

———. "The Growing Role of Minerals and Metals for a Low Carbon Future." June 2017. http://documents.worldbank.org/curated/en/207371500386458722/The-Growing-Role-of-Minerals-and-Metals-for-a-Low-Carbon-Future.

World Economic Forum. "The Fourth Industrial Revolution." www.weforum.org.

World Food Programme. "New WFP Report Shows Access to Food Grossly Unequal As Coronavirus Adds to Challenges." October 16, 2020. https://www.wfp.org/news/new-wfp-report-shows-access-food-grossly-unequal-coronavirus-adds-challenges.

World Health Organization. "As More Go Hungry and Malnutrition Persists, Achieving Zero Hunger by 2030 in Doubt, UN Report Warns." July 13, 2020. https://www.who.int/news/item/13-07-2020-as-more-go-hungry-and-malnutrition-persists-achieving-zero-hunger-by-2030-in-doubt-un-report-warns.

World Rainforest Management. "The Conservation Industry's Agenda in Times of Crisis." *WRM Bulletin* 249 (May 14, 2020). https://www.wrm.org.uy/bulletin-articles/the-conservation-industrys-agenda-in-times-of-crisis.

World Resources Institute. "South Africa: Strong Foundations for a Just Transition." December 23, 2021. https://www.wri.org/update/south-africa-strong-foundations-just-transition.

Xu, Chi, et al. "Future of the Human Climate Niche." *Proceedings of the Na-*

*tional Academy of Sciences* 117, no. 21 (May 2020): 11350–55. https://doi. org/10.1073/pnas.1910114117.

Yuen, Eddie, Daniel Burton-Rose, and George Katsiaficas, eds. *Confronting Capitalism: Dispatches from a Global Movement.* Brooklyn, NY: Soft Skull Press, 2004.

Zapatista. *Zapatista Encuentro: Documents from the 1996 Encounter for Humanity and Against Neoliberalism.* New York: Seven Stories, 2002.

Zibechi, Raúl. *Dispersing Power Social Movements as Anti-state Forces.* Oakland, CA: AK Press, 2010.

# NOTES

## Introduction: A Global People's Movement

1. Nick Buxton, in discussion with the author, December 2019.
2. Taking place just five months after the "Battle of Seattle" protests against the World Trade Organization, the Water Wars were sparked by the World Bank's refusal to renew a $25 million loan to Bolivia unless it privatized its municipal water supplies. A coalition of peasants, workers, urban families, and coca growers rose up in 2000 against government plans to comply with the World Bank's edict, declaring, "The water is ours, damn it!" The movement dug up and blockaded roads in Cochabamba to keep the police and military from putting down the uprising. Resistance was coordinated by people's assemblies, who debated alternatives to water privatization. See Oscar Olivera and Tom Lewis, ¡Cochabamba! Water War in Bolivia (Cambridge, MA: South End Press, 2004).
3. Rishika Pardikar, "Global North Is Responsible for 92% of Excess Carbon Emissions," Eos, October 28, 2020, https://eos.org/articles/global-north-is-responsible-for-92-of-excess-emissions.
4. World Bank, "$CO_2$ Emissions (Metric Tons per Capita)," https://data.worldbank.org/indicator/EN.ATM.CO2E.PC.
5. On the history of Barack Obama's embrace of the Green New Deal, see Michael Grunwald, The New New Deal: The Hidden Story of Change in the Obama Era (New York: Simon & Schuster, 2012). For the background on US negotiations at Copenhagen, see David Adam, "US Planning to Weaken Copenhagen Climate Deal, Europe Warns," The Guardian, September 15, 2009; and Suzanne Goldenberg and Allegra Stratton, "Barack Obama's Speech Disappoints and Fuels Frustration at Copenhagen," The Guardian, December 18, 2009. For an assessment of Obama's flawed legacy on climate change, see Marianne Lavelle, "2016: Obama's Climate Legacy Marked by Triumphs and Lost Opportunities," Inside Climate News, December 26, 2016.

6.  Building Bridges Collective, *Space for Movement?: Reflections from Bolivia on Climate Justice, Social Movements, and the State* (Leeds, UK: Footprint Workers Co-op, 2010), 8.

7.  Adam Weltz, "Lumumba Di-Aping: 'We Have Been Asked to Sign a Suicide Pact,'" *Reimagine: Movements Making Media*, December 14, 2009, https://www.reimaginerpe.org/cj/news/12-9.

8.  Evo Morales Ayma, "We in the Social Movements Know the Problems and Also the Solutions," in *The Earth Does Not Belong to Us, We Belong to the Earth* (Bolivian Ministry of Foreign Relations, 2010), 110.

9.  Indigenous Environmental Network, "Exposing REDD: The False Climate Solution," October 22, 2012.

10. Indigenous Environmental Network, "Carbon Offsets Cause Conflict and Colonialism," May 18, 2016.

11. Ramachandra Guha and Juan Martinez-Alier, *Varieties of Environmentalism: Essays North and South* (London: Earthscan, 1997), xxi. See also Juan Martinez-Alier, *The Environmentalism of the Poor: A Study of Ecological Conflicts and Valuation* (Northampton, MA: Edward Elgar, 2002). Another key influence of this book is Rob Nixon's *Slow Violence and the Environmentalism of the Poor* (Cambridge, MA: Harvard University Press, 2011), which, as its title suggests, draws extensively on Guha and Martinez-Alier's work.

12. Guha and Martinez-Alier, *Varieties of Environmentalism*, xxi.

13. In addition to the work of Guha and Martinez-Alier, my concept of *environmentalism from below* draws on the rich legacy of historiography of the poor and peasantry that grew from the work of the radical British historian E. P. Thompson, including the important tradition of Subaltern Studies in India.

14. I am drawing a parallel here between environmentalism from below and the autonomous agency of the working class, as theorized, for example, by Antonio Negri in *Marx beyond Marx* (New York: Pluto Press, 1992).

15. Benedickt Bruckner et al., "Impacts of Poverty Alleviation on National and Global Carbon Emissions," *Nature Sustainability* 5 (2022): 311–20.

16. Lucas Chancel, "Global Carbon Inequality over 1990–2019," *Nature Sustainability* 5 (2022): 931–38.

17. On the resistance at Standing Rock, see Nick Estes, *Our History is the Future: Standing Rock Versus the Dakota Access Pipeline, and the Long Tradition of Indigenous Resistance* (New York: Verso Books, 2019) and Jaskiran Dhillon and Nick Estes, eds., *Standing with Standing Rock: Voices from the #NoDALP Movement* (Minneapolis: University of Minnesota Press, 2019).

18. For a discussion of experiments in popular sovereignty across the Pink Tide nations, see Gianpaolo Baiocchi, *We, the Sovereign* (Medford, MA:

Polity Press, 2018).

19. Zapatista, *Zapatista Encuentro: Documents from the 1996 Encounter for Humanity and Against Neoliberalism* (New York: Seven Stories Press, 2002).

20. See Raúl Zibechi, *Dispersing Power: Social Movements as Anti-state Forces* (Oakland: AK Press, 2012).

21. See Deborah McGregor, Steven Whitaker, and Mahisha Sritharan, "Indigenous Environmental Justice and Sustainability," *Current Opinion in Environmental Sustainability* 43 (2020): 35–40, https://doi.org/10.1016/j.cosust.2020.01.007.

22. Cormac Cullinan, *Wild Law: A Manifesto for Earth Justice* (New York: Chelsea Green, 2011).

23. McGregor, Whitaker, and Sritharan, "Indigenous Environmental Justice and Sustainability," 38.

24. Emily Achtenberg, "Contested Development: The Geopolitics of Bolivia's TIPNIS Conflict," *NACLA Report on the Americas* 46, no. 2 (August 2013): 6–11.

25. Both are available at World People's Conference on Climate Change and the Rights of Mother Earth, available at https://pwcc.wordpress.com/.

26. World People's Conference on Climate Change and the Rights of Mother Earth, *People's Agreement of Cochabamba*, April 22, available at https://pwcc.wordpress.com/.

27. In 1916, Luxemburg wrote, "Bourgeois society stands at the crossroads, either transition to Socialism or regression into Barbarism," in the *Junius Pamphlet* (1916), Marxist Internet Archive, https://www.marxists.org/archive/luxemburg/1915/junius.

28. Although the term "decolonial ecologies" is my own coinage, it draws on an important tradition of theorization based in Latin America in particular. For an overview of this tradition, see Walter D. Mignolo and Catherine E. Walsh, *On Decoloniality: Concepts, Analytics, Praxis* (Durham, NC: Duke University Press, 2018).

29. Naomi Klein, *This Changes Everything: Capitalism versus the Climate* (New York: Simon & Schuster, 2014), 412.

30. Vladimir Lenin, *Imperialism: The Highest Stage of Capitalism* (1917; repr., New York: Penguin Classics, 2010).

31. Harpreet Kaur Paul and Dalia Gebrial, eds., *Perspectives on a Global Green New Deal* (New York: Rosa Luxemburg Foundation, 2021), 9.

32. Andrew Simms, et al., *A Green New Deal: Joined-Up Policies to Solve the Triple Crunch of the Credit Crisis, Climate Change, and High Oil Prices*, New Economics Foundation, July 2008, https://neweconomics.org/2008/07/green-new-deal.

33. Simms et al., *A Green New Deal.*

34. Edward Barbier, *A Global Green New Deal: Rethinking the Economic Recovery* (United Nations Environmental Program, 2009).

35. Ashley Dawson, "Why We Need a Global Green New Deal," *New Politics* 12, no. 4 (Winter 2010).

36. Frances Ryan, "For the Tories, Eliminating the Deficit Was Just a Pretext to Slash the State," *The Guardian*, March 8, 2018.

37. Van Jones, "Working Together for a Green New Deal" *The Nation*, October 29, 2008, https://www.thenation.com/article/archive/working-together-green-new-deal/.

38. See, for example, Jonathan Chait, "Obama Had a Green New Deal, and It Worked. Let's Do That Again," *New York Magazine*, April 26, 2019, https://nymag.com/intelligencer/2019/04/obamas-green-new-deal-worked-climate-change.html.

39. Sarah Knuth, "Rentiers of the Low-Carbon Economy? Renewable Energy's Extractive Fiscal Geographies," *Environment and Planning A: Economy and Space* (2021): 1–17.

40. Alexander Sammon, "How the Bank Bailout Hobbled the Climate Fight," *New Republic*, October 22, 2018.

41. David Wallace-Wells, "Hardly Anyone Talks about How Fracking Was an Extraordinary Boondoggle," *New York Times*, July 27, 2022.

42. John Muyskens and Juliet Eilperin, "Biden Calls for 100 Percent Clean Electricity By 2035. Here's How Far We Have to Go," *Washington Post*, July 30, 2020.

43. Oliver Milman, "A Closer Look at Joe Manchin's Ties to the Fossil Fuel Industry," *Mother Jones*, October 21, 2021.

44. For a full overview of the IRA, see Rebecca Lerner, "The Senate Just Passed One of the Biggest Bills to Fight Climate Change, Ever," *Vox*, August 7, 2022, https://www.vox.com/policy-and-politics/2022/7/28/23281757/whats-in-climate-bill-inflation-reduction-act.

45. Quoted in Nina Lakhani, "Landmark US Climate Bill Will Do More Harm Than Good, Groups Say," *The Guardian*, August 9, 2022.

46. Paul Krugman, "Did the Democrats Just Save Civilization?" *New York Times*, August 8, 2022.

47. Chelsea Harvey, "Climate Pledges Still Not Enough to Keep Warming below 2-Degree Limit," *Scientific American*, November 23, 2021, https://www.scientificamerican.com/article/climate-pledges-still-not-enough-to-keep-warming-below-2-degree-limit/.

48. Quoted in Lakhani, "Landmark US Climate Bill."

49. Oliver Milman, "Biden's Approval of Willow Project Shows Inconsistency

of US's First 'Climate President,'" *The Guardian*, March 14, 2023.

50. Asad Rehman, "The 'Green New Deal' Supported by Ocasio-Cortez and Corbyn Is Just a New Form of Colonialism," *The Independent*, May 4, 2019.

51. Coalition of Climate Justice Movements, "COP25, Social Movements and Climate Justice," *The Ecologist*, December 2, 2019.

52. For a damning statistical analysis of what academics call "ecologically unequal exchange" and others call "plunder," see Andreas Mayer and Willi Haas, "Cumulative Material Flows Provide Indicators to Quantify the Ecological Debt," *Journal of Political Ecology* 23 (2016): 350–63.

53. Kate Aronoff, "The Socialist Win in Bolivia and the New Era of Lithium Extraction," *New Republic*, October 19, 2020.

54. Statistics from Nick Ferris, "Why Critical Mineral Supplies Won't Scupper the Energy Transition," *Energy Monitor*, June 23, 2021, https://www.energymonitor.ai/finance/risk-management/why-we-need-a-level-headed-approach-to-energy-transition-minerals.

55. World Bank, "The Growing Role of Minerals and Metals for a Low Carbon Future," June 2017, https://documents.worldbank.org/en/publication/documents-reports/documentdetail/207371500386458722/the-growing-role-of-minerals-and-metals-for-a-low-carbon-future.

56. Nafeez Ahmed, "We Don't Mine Enough Rare Earth Metals to Replace Fossil Fuels with Renewable Energy," *Vice*, December 12, 2018, https://www.vice.com/en/article/a3mavb/we-dont-mine-enough-rare-earth-metals-to-replace-fossil-fuels-with-renewable-energy.

57. US Geological Survey, *Mineral Commodity Summaries 2021* (Reston, VA: US Geological Survey, 2021), https://pubs.usgs.gov/periodicals/mcs2021/mcs2021.pdf.

58. Ferris, "Why Critical Mineral Supplies."

59. Lizabeth Cohen, *A Consumers' Republic: The Politics of Mass Consumption in Postwar America* (New York: Knopf, 2003).

60. Naomi Klein, *On Fire: The (Burning) Case for a Green New Deal* (New York: Simon & Schuster, 2019), 264.

61. Klein, *On Fire*, 264.

62. Ulrich Brand and Markus Wissen, "The Imperial Mode of Living," in Clive L. Spash, ed., *The Routledge Handbook of Ecological Economics: Nature and Society* (New York: Routledge, 2017). 152–61.

63. George Monbiot, "How Labour Could Lead the Global Economy Out of the 20th Century," *The Guardian*, October 11, 2017.

64. Teppo Eskelinen, "Possibilities and Limits of Green Keynesianism," in Kasja Borgnäs et al., eds., *The Politics of Eco-Socialism: Transforming Welfare* (New York: Oxford: Routledge, 2015), 101–15.

65. See, for example, Nicholas Stern, *The Economics of Climate Change: The Stern Review* (New York: Cambridge University Press, 2007); and *An Ecomodernist Manifesto*.

66. Gareth Dale, Manu V. Mathai, and Jose A. Puppim De Olivera, eds., *Green Growth: Ideology, Political Economy and the Alternatives* (London: Zed Books, 2016), 9.

67. Fridolin Krausmann et al., "Growth in Global Materials Use, GDP and Population during the 20th Century," *Ecological Economics* 68, no. 10 (2009): 2696–705.

68. Stefan Giljum et al., "Global Patterns of Material Flows and Their Socio-economic and Environmental Implications: A MFA Study on All Countries World-Wide from 1980 to 2009," *Resources* 3, no. 1 (2014): 319–39.

69. Jason Hickel, "A Response to Noah Smith about Global Poverty," August 6, 2019, https://www.jasonhickel.org/blog.

70. Indigenous Environmental Network, "Just Transition," https://www.ienearth.org/justtransition/.

71. Mahmoud Salari, Roxana J. Javid, and Hamid Noghanibehambari, "The Nexus between $CO_2$ Emissions, Energy Consumption, and Economic Growth in the US," *Economic Analysis and Policy* 69 (March 2021): 182–94.

72. On the racism and imperialism of the New Deal, see Justin Podur, "Leaving Behind the Racist and Imperialist Baggage of the Original New Deal," in Kaur Paul and Gebrial, eds., *Perspectives on a Global Green New Deal*, 123–26. Andreas Malm has suggested that war communism during the early years of the Soviet Union is a better model for the future, but this is obviously a hard sell in the US given long-standing traditions of red-baiting. See Andreas Malm, *Corona, Climate, and Chronic Emergency: War Communism in the 21st Century* (New York: Verso Books, 2020).

73. Leah Temper et al., "Movements Shaping Climate Futures: A Systematic Mapping of Protests Against Fossil Fuel and Low-Carbon Energy Projects," *Environmental Research Letters* 15 (2020), https://doi.org/10.1088/1748-9326/abc197.

# Chapter 1: Decolonizing Food

1. Mujib Mashal, Emily Schmall, and Hari Kumar, "As Angry Farmers Take to New Delhi's Streets, Protests Turn Violent," *New York Times*, January 25, 2021.

2. Zafar Imran, "Climate Change in the Indian Farmers' Protest," *Le Monde diplomatique*, February 1, 2021, https://mondediplo.com/outsidein/climate-indian-farmers.

3. Omair Ahmad, "Climate Crisis Is Foundation of Indian Farmers' Protests," The Third Pole, January 25, 2021, https://www.thethirdpole.net/en/food/climate-crisis-is-foundation-of-indian-farmers-protests/.

4. United Nations, "The State of Food Security and Nutrition in the World 2021," https://www.fao.org/3/cb4474en/online/cb4474en.html.

5. Food Research and Action Center, "Hunger and Poverty in America," https://frac.org/hunger-poverty-america.

6. CARE Food and Water Systems, Left Out and Left Behind: Ignoring Women Will Prevent Us from Solving the Hunger Crisis, November 2020, https://care.org/wp-content/uploads/2020/08/Left-Out-and-Left-Behind.pdf.

7. World Food Programme, "New WFP Report Shows Access to Food Grossly Unequal As Coronavirus Adds to Challenges," October 16, 2020, https://www.wfp.org/news/new-wfp-report-shows-access-food-grossly-unequal-coronavirus-adds-challenges.

8. Committee on World Food Security, "Impacts of COVID-19 on Food Security and Nutrition: Developing Effective Policy Responses to Address the Hunger and Malnutrition Epidemic," September 2020, https://www.fao.org/3/cb1000en/cb1000en.pdf.

9. William G. Moseley and Jane Battersby, "The Vulnerability and Resilience of African Food Systems, Food Security, and Nutrition in the Context of the COVID-19 Pandemic," African Studies Review 63, no. 3 (2020): 1–13.

10. International Food Policy Research Institute, "Global Food Prices Remain High, Food Security Remains Uncertain This Year," January 13, 2023, https://www.ifpri.org/news-release/global-food-prices-remain-high-food-security-remains-uncertain-year-epoch-times.

11. "Keeping Things Cornucopious: The World's Food System Has So Far Weathered the Challenge of COVID-19," The Economist, May 9, 2020.

12. Jan Douwe van der Ploeg, "Growing Back Stronger: Choosing Resilient Food Systems in the Wake of COVID-19," June 2020, https://www.tni.org/en/foodsystems.

13. Sir Robert Watson, quoted in "Media Release: Nature Dangerous Decline 'Unprecedented'; Species Extinction Rates 'Accelerating,'" IPBES, May 7, 2019.

14. Watson, "Nature Dangerous Decline."

15. Watson, "Nature Dangerous Decline."

16. The term ecocide was coined by American biologist Arthur Galston, who spoke out against the massive environmental damage caused by the use of the defoliant chemical Agent Orange by US military forces in Vietnam during the 1960s and 1970s. Richard Falk drafted an Ecocide Convention

in 1973. The term has recently gained prominence in efforts to criminalize human activities that cause extensive damage to, destruction of, or loss of ecosystems in a given territory, and which diminish the health and well-being of species within these ecosystems (including humans).

17. GRAIN, *The Great Climate Robbery: How the Food System Drives Climate Change and What We Can Do about It* (Oxford: New Internationalist Publications, 2015).

18. The most important exception to this rule is the Red Nation's *Red Deal*, which includes demands for a decolonial agriculture. Also worth noting (although it is still focused exclusively on the US) is Raj Patel and Jim Goodman's "A Green New Deal for Agriculture," *Jacobin*, April 4, 2019.

19. Eric Holt-Giménez, *A Foodie's Guide to Capitalism: Understanding the Political Economy of What We Eat* (New York: NYU Press, 2017), 38.

20. Sophia Murphy, David Burch, and Jennifer Clapp, *Cereal Secrets: The World's Largest Grain Traders and Global Agriculture*, Oxfam Research Reports, August 2012), https://www.oxfam.org/en/research/cereal-secrets-worlds-largest-grain-traders-and-global-agriculture.

21. David Harvey, *The New Imperialism* (New York: Oxford University Press, 2003), 67.

22. George Monbiot, *Regenesis: Feeding the World without Devouring the Planet* (New York: Penguin, 2023), 35.

23. John Bellamy Foster and Intan Suwandi, "COVID-19 and Catastrophe Capitalism," *Monthly Review*, June 1, 2020.

24. Brian Ashley et al., *One Million Climate Jobs: Moving South Africa Forward on a Low-Carbon, Wage-Led and Sustainable Path*, Alternative Information and Development Centre, December 2016.

25. Peter Schwartzman and David Schwartzman, *The Earth Is Not for Sale: A Path out of Fossil Capitalism to the Other World That Is Still Possible* (Hackensack, NJ: World Scientific, 2019), 120.

26. Schwartzman and Schwartzman, *Earth Is Not for Sale*, 72.

27. Vandana Shiva, *Who Really Feeds the World* (Berkeley, CA: Zed Books, 2015), 29.

28. David Montgomery, *Dirt: The Erosion of Civilization* (Berkeley: University of California Press, 2007), 6.

29. Chris Arsenault, "Only 60 Years of Farming Left If Soil Degradation Continues," *Scientific American*, December 5, 2014.

30. Ian Bogost, "Can You Sue a Robocar?" *The Atlantic*, March 20, 2018.

31. Bogost, "Can You Sue a Robocar?"

32. Norman Mayersohn, "How High Tech Is Transforming One of the Oldest Jobs: Farming," *New York Times*, September 6, 2019.

33. Karl Marx, *Capital*, Vol. 1, trans. Ben Fowkes (London: Vintage, 1976), 502. For extended discussion of the trend toward automation, see Moise Postone, *Time, Labor, and Social Domination: A Reinterpretation of Marx's Critical Theory* (New York: Cambridge University Press, 1996), 338.

34. On the tendency to increase the organic composition of capital, and how this tendency has affected all aspects of mining, see Martín Arboleda, *Planetary Mine: Territories of Extraction under Late Capitalism* (New York: Verso Books, 2020), 2–5.

35. See David Harvey, "Rate and Mass," *New Left Review* 130 (July/August 2021), https://newleftreview.org/issues/ii130/articles/david-harvey-rates-and-mass.

36. Jamie Morgan, "The Fourth Industrial Revolution Could Lead to a Dark Future," *The Conversation*, January 9, 2020.

37. World Economic Forum, "The Fourth Industrial Revolution," www.weforum.org.

38. Knvul Sheikh, "A Growing Presence on the Farm: Robots." *New York Times*, February 13, 2020.

39. Eric Niiler, "Why Gene Editing Is the Next Food Revolution," *National Geographic*, August 10, 2018.

40. Quoted in Vandana Shiva, Prerna Anilkumar, and Urvee Ahluwalia, *Ag One: The Recolonization of Agriculture* (Navdanya/RESTE, 2020).

41. Quoted in "Egyptians Riot over Bread Crisis," *The Telegraph*, April 8, 2008.

42. "Rising Food Prices Can Topple Governments, Too" NPR, January 30, 2011.

43. Andreas Malm, "Tahrir Submerged? Five Theses on Revolution in the Age of Climate Change," *Capitalism Nature Socialism* 25, no. 3 (March 2014): 28–44, https://doi.org/10.1080/10455752.2014.891629.

44. World Economic Forum, "The Fourth Industrial Revolution Is Changing How We Grow, Buy, and Choose What We Eat," August 3, 2018, https://www.weforum.org/agenda/2018/08/the-fourth-industrial-revolution-is-changing-how-we-grow-buy-and-choose-what-we-eat.

45. Aaron Bastani, *Fully Automated Luxury Communism: A Manifesto* (New York: Verso Books, 2019), 237.

46. Out of the Woods Collective, *Hope Against Hope: Writings on Ecological Crisis* (Philadelphia: Common Notions, 2020), 115.

47. On the history of imperial science, see Carolyn Merchant, *The Death of Nature: Women, Ecology, and the Scientific Revolution* (New York: HarperOne, 1990).

48. Shiva, Anilkumar, and Ahluwalia, *Ag One*.

49. Robert Biel, *Sustainable Food Systems: The Role of the City* (London: University College London Press, 2016), 12.

50. Marx, *Capital*, Vol. 1, 638.

51. David R. Montgomery, "Peak Soil," *New Internationalist*, December 2, 2008.

52. Holt-Giménez, *Foodie's Guide*, 46.

53. Holt-Giménez, *Foodie's Guide*, 46.

54. Cited in "Public Law 480: 'Better Than a Bomber,'" *Middle East Report* 145 (March/April 1987).

55. Mike Davis, *Late Victorian Holocausts: El Niño Famines and the Making of the Third World* (New York: Verso Books, 2017).

56. Holt-Giménez, *Foodie's Guide*, 47.

57. Daniel Zwerdling, "'Green Revolution' Trapping India's Farmers in Debt," NPR, April 14, 2009.

58. Mike Davis, *Planet of Slums* (New York: Verso Books, 2006).

59. Holt-Giménez, *Foodie's Guide*, 48.

60. Boyce Rensberger, "Experts Asks Action to Avoid Millions of Deaths in Food Crisis," *New York Times*, July 26, 1974.

61. Jennifer Clapp and Walter G. Moseley, "This Food Crisis Is Different: COVID-19 and the Fragility of the Neoliberal Food Order," *Journal of Peasant Studies* 47, no. 7 (2020): 1393–417.

62. "Farm Aid: Saving the Family Farm," NPR, November 23, 2006.

63. Eric Toussaint, "The Mexican Debt Crisis and the World Bank," *Global Research*, August 8, 2019, https://www.globalresearch.ca/mexican-debt-crisis-world-bank/5685881.

64. Giovanni Arrighi, "The Social and Political Economy of Global Turbulence," *New Left Review* 20 (2003): 5–71.

65. John Madeley, *Hungry for Trade: How the Poor Pay for Free Trade* (London: Zed, 2000), 75.

66. Raj Patel and Philip McMichael, "A Political Economy of the Food Riot," *Review: A Journal of the Fernand Braudel Center* 32, no. 1 (2009): 9–35.

67. Imad A. Moosa and Nisreen Moosa, "The IMF as an Instigator of Riots and Civil Unrest," in *Eliminating the IMF: An Analysis of the Debate to Keep, Reform or Abolish the Fund* (Cham, Switzerland: Palgrave Macmillan, 2019), 55–87.

68. Vijay Prashad, "The IMF Is Utterly Indifferent to the Pain It's Causing," *Truthdig*, October 15, 2019, https://www.truthdig.com/articles/the-imf-is-utterly-indifferent-to-the-pain-its-causing/.

69. For an overview of the Global Justice Movement, see Eddie Yuen, Daniel Burton-Rose, and George Katsiaficas, eds., *Confronting Capitalism: Dispatches from a Global Movement* (Brooklyn, NY: Soft Skull Press, 2004).

70. UN Food and Agriculture Organization, "Rome Declaration on World

Food Security," November 13–17, 1996.

71. Melinda Cooper, *Life as Surplus: Biotechnology and Capitalism in the Neoliberal Era* (Seattle: University of Washington Press, 2008), 19.

72. Cooper, *Life as Surplus*, 21.

73. Anthony D. So et al., "Is Bayh-Dole Good for Developing Countries? Lessons from the US Experience," *PLoS Biology* 6, no. 10 (October 2008), https://doi.org/10.1371/journal.pbio.0060262.

74. Cooper, *Life as Surplus*, 23.

75. Biel, *Sustainable Food Systems*, 8; and Douglas Puffert, "Path Dependence," Economic History Association Encyclopedia, February 10, 2008, www. eh.net.

76. Daniel Cressey, "Widely Used Herbicide Linked to Cancer," *Nature News*, March 24, 2015, https://doi.org/10.1038/nature.2015.17181.

77. Jordan Wilkerson, "Why Roundup Ready Crops Have Lost Their Allure," *Science in the News*, August 10, 2015.

78. Laura Carlsen, "WTO Kills Farmers: In Memory of Lee Kyung Hae," Counter Currents, September 16, 2003, http://www.countercurrents.org/glocarlsen160903.htm.

79. Rajeev Patel, "International Agrarian Restructuring and the Practical Ethics of Peasant Movement Solidarity," *Journal of African and Asian Studies* 41, no. 1-2 (2006): 71–93.

80. La Via Campesina, "Globalize Hope: New Film on the History of La Via Campesina," 2020, https://tv.viacampesina.org/Globalize-Hope?lang=en.

81. La Via Campesina, "Globalize Hope."

82. La Via Campesina, "Globalize Hope."

83. Annette Aurélie Desmarais, *La Vía Campesina: Globalization and the Power of Peasants* (Ann Arbor, MI: Pluto, 2007), 34.

84. La Via Campesina, "International Conference of Agrarian Reform: Marabá Declaration: Via Campesina," April 22, 2016, https://viacampesina.org/en/international-conference-of-agrarian-reform-declaration-of-maraba1/.

85. Desmarais, *La Vía Campesina*, 25.

86. La Via Campesina, "Globalize Hope."

87. Stefano Liberti, *Land Grabbing: Journeys in the New Colonialism* (New York: Verso Books, 2013); Marc Edelman et al., eds., *Global Land Grabbing and Political Reactions "from Below"* (New York: Routledge, 2018); Transnational Institute, *Rogue Capitalism and the Financialization of Territories and Nature*, September 2020, www.tni.org.

88. Alex Wijeratna, *Agroecology: Scaling-Up, Scaling-Out*, ActionAid International, April 2018, https://actionaid.org/publications/2018/agroecology-scaling-scaling-out.

89. James C. Scott, *Seeing Like a State: How Certain Schemes to Improve the Human Condition Have Failed* (New Haven, CT: Yale University Press, 1999).

90. Global Agriculture, "Study: Agroecology Can Help to Mitigate Climate Change," April 18, 2019, www.globalagriculture.org.

91. Vandana Shiva, *Who Really Feeds the World? The Failures of Agribusiness and the Promise of Agroecology* (Berkeley: Zed Books, 2015), 56.

92. Shiva, *Who Really Feeds the World?* 54.

93. Shiva, *Who Really Feeds the World?* 57.

94. David R. Montgomery, *Growing a Revolution: Bringing Our Soil Back to Life* (New York: W.W. Norton, 2017), 48.

95. Shiva, *Who Really Feeds the World?* 2.

96. Shiva, *Who Really Feeds the World?* 31.

97. Montgomery, *Growing a Revolution*, 47.

98. Montgomery, *Growing a Revolution*, 48.

99. K. R. Shivanna, Rajesh Tandon, and Monika Koul, "'Global Pollinator Crisis' and Its Impact on Crop Productivity and Sustenance of Plant Diversity," in *Reproductive Ecology of Flowering Plants: Patterns and Processes*, ed. Rajesh Tandon, K. R. Shivanna, and Monika Koul (Singapore: Springer, 2020): 395–413.

100. Montgomery, *Growing a Revolution*, 48; and Shiva, *Who Really Feeds the World?* 34.

101. Wijeratna, *Agroecology*.

102. Wijeratna, *Agroecology*, 6.

103. Montgomery, *Growing a Revolution*, 283.

104. Giovanni Tamburini et al., "Agricultural Diversification Promotes Multiple Ecosystem Services without Compromising Yield," *Science Advances* 6, no. 45 (November 2020), https://doi.org/10.1126/sciadv.aba1715.

105. Montgomery, *Growing a Revolution*, 50.

106. Ivette Perfecto, John Vandermeer, and Angus Wright, *Nature's Matrix: Linking Agriculture, Biodiversity Conservation, and Food Sovereignty* (New York: Routledge, 2019), 4.

107. Perfecto, Vandermeer, and Wright, *Nature's Matrix*, 14.

108. Cited in Montgomery, *Growing a Revolution*, 226.

109. Cited in Montgomery, *Growing a Revolution*, 226.

110. GRAIN, *Great Climate Robbery*.

111. Biel, *Sustainable Food Systems*, 62.

112. Antonio Negri, *Marx beyond Marx: Lessons on the Grundrisse* (New York: Autonomedia, 1991).

113. Andreas Malm, *The Progress of This Storm: Nature and Society in a Warming World* (New York: Verso Books, 2017), 197–207.

114. Veronique Dupont, "US 'Superweeds' Epidemic Shines Spotlight on GMOs," *Phys*, January 13, 2014, https://phys.org/news/2014-01-super-weeds-epidemic-spotlight-gmos.html.

115. Vincent Ricciardi et al., "Higher Yields and More Biodiversity on Smaller Farms," *Nature Sustainability* (2021), https://doi.org/10.1038/s41893-021-00699-2.

116. Biel, *Sustainable Food Systems*, 88.

117. For a critique of "sustainable intensification," see Friends of the Earth International, the Transnational Institute, and Crocevia, *Junk Agroecology: The Corporate Capture of Agroecology for a Partial Ecological Transition without Social Justice*, April 2020; and Holt-Giménez, *Foodie's Guide*, 180–82.

118. La Via Campesina, "Opinion: Agroecology for Gender Equality: Via Campesina," Via Campesina English, September 20, 2016, https://viacampesina.org/en/opinion-agroecology-for-gender-equality/.

119. Cyndie Shearing "Women Count in Agriculture," American Farm Bureau Federation, May 1, 2019, https://www.fb.org/viewpoints/women-count-in-agriculture.

120. Cited in Márcia Maria Tait Lima and Vanessa Brito de Jesus, "Questions about Gender and Technology in Agroecology," *Scientiae Studia* 15, no. 1 (November 2018): 73–96, https://doi.org/10.11606/51678-31662017000100005.

121. Lima and Brito de Jesus, "Questions about Gender."

122. Iridiani Graciele Seibert et al., "Without Feminism, There Is No Agroecology," Civil Society and Indigenous Peoples' Mechanism, August 2019, http://www.csm4cfs.org/wp-content/uploads/2019/10/CSM-Agroecology-and-Feminism-September-2019_compressed.pdf.

123. Seibert et al., "Without Feminism."

124. CARE, *Left Out and Left Behind*.

125. Seibert et al., "Without Feminism."

126. Seibert et al., "Without Feminism."

127. Seibert et al., "Without Feminism."

128. Shiva, *Who Really Feeds the World?* 128.

129. Sônia Fátima Schwendler and Lucia Amaranta Thompson, "An Education in Gender and Agroecology in Brazil's Landless Rural Workers' Movement," *Gender and Education* 29, no. 1 (January 2017): 100–114, https://doi.org/10.1080/09540253.2016.1221596.

130. La Via Campesina, "Struggles of La Via Campesina | For Agrarian Reform and the Defense of Life, Land and Territories," Via Campesina English, October 16, 2017, https://viacampesina.org/en/struggles-la-via-campesina-agrarian-reform-defense-life-land-territories/.

131. Wijeratna, *Agroecology.*

132. This list of climate finance measures for rural areas is derived from Ashley et al., *One Million Climate Jobs.*

133. "Rogue Capitalism and the Financialization of Territories and Nature," Transnational Institute, September 28, 2020, https://www.tni.org/en/rogue-capitalism.

134. La Via Campesina, "Struggles of La Via Campesina."

135. Shiney Varghese, "Indian Farmers Score Major Victory after Year-Long Strike," *Common Dreams*, December 8, 2021, https://www.common-dreams.org/views/2021/12/08/indian-farmers-score-major-victory-after-year-long-strike.

136. Institute for Agriculture and Trade Policy, "We Stand with India's Farmers! Now Let's Connect the Dots between the Forces of Neoliberalism That Stifle Farmers, from India to the U.S.," February 20, 2021, https://www.iatp.org/sites/default/files/2021-02/Solidarity%20Statement%20for%20Indias%20farmers%20from%20US%20based%20food%20and%20farm%20justice%20organisations%20February%202021.pdf.

137. La Via Campesina, "Struggles of La Via Campesina."

138. See Ashley Dawson, *Extreme Cities: The Peril and Promise of Urban Life in the Age of Climate Change* (New York: Verso Books, 2017).

## Chapter 2: Urban Climate Insurgency

1.  My account of the Chilean uprising is based on Romina A. Green Rioja, "'Until Living Becomes Worth It': Notes from the Chilean Uprising," *Abusable Past* (blog), November 1, 2019, https://www.radicalhistoryreview.org/abusablepast/until-living-becomes-worth-it-notes-from-the-chilean-uprising/.

2.  Rioja, "Until Living Becomes Worth It."

3.  Hot City Collective, "Hot City: Compound Crisis and Popular Struggle in NYC," *Verso Books* (blog), August 3, 2020, https://www.versobooks.com/blogs/news/4811-hot-city-compound-crisis-and-popular-struggle-in-nyc.

4.  From the synopsis of the book by Evan Calder Williams, *Combined and Uneven Apocalypse: Luciferian Marxism* (Winchester, UK: Zero Books). The synopsis is available on the Zero Books website: https://www.johnhunt-publishing.com/zer0-books/our-books/combined-and-uneven-apocalypse.

5.  Andreas Malm, "Tahrir Submerged? Five Theses on Revolution in the Age of Climate Change," *Capitalism Nature Socialism* 25, no. 3 (March 2014): 28–44, https://doi.org/10.1080/10455752.2014.891629.

6.  Brendan Gleeson, *The Urban Condition* (New York: Routledge, 2014).

7. UN Department of Economic and Social Affairs, "68% of World Population Projected to Live in Urban Areas by 2050, Says UN," May 16, 2018, https://www.un.org/development/desa/en/news/population/2018-revision-of-world-urbanization-prospects.html.

8. UN-Habitat, *Slum Almanac 2015–2016*, https://unhabitat.org/slum-almanac-2015-2016-0.

9. Jennifer Robinson, *Ordinary Cities: Between Modernity and Development* (New York: Routledge, 2006), 1.

10. UN-Habitat, "Urban Indicators Database," https://data.unhabitat.org/pages/housing-slums-and-informal-settlements

11. Mihir Zaveri et al., "How the Storm Turned Basement Apartments into Death Traps," *New York Times*, September 2, 2021.

12. For a full discussion of the history of these austerity policies and their link to urbanization in the Global South, see Mike Davis, *Planet of Slums* (New York: Verso Books, 2006).

13. James Holston, *Insurgent Citizenship: Disjunctions of Democracy and Modernity in Brazil* (Princeton, NJ: Princeton University Press, 2007).

14. Holston, *Insurgent Citizenship*, 246.

15. The literature tracking this opposition is voluminous, including work as diverse as Raymond Williams, *The Country and the City* (New York: Oxford University Press, 1973); and Maria Kaika, *City of Flows: Modernity, Nature, and the City* (New York: Routledge, 2004).

16. James W. Moore, *Anthropocene or Capitalocene?: Nature, History, and the Crisis of Capitalism* (Oakland, CA: PM Press/Kairos, 2016).

17. United Nations, "Sustainable Development Goal #11," https://sdgs.un.org/goals/goal12.

18. For a survey of radical design initiatives in the cities of the Global South, see Cynthia E. Smith, *Design with the Other 90%: Cities* (New York: Cooper-Hewitt, National Design Museum, 2007).

19. Davis, *Planet of Slums*, 121–50.

20. On the environmental hazards of what he calls "slum ecologies," see Davis, *Planet of Slums*, 121–50.

21. I developed this term in collaboration with the Occupy Climate Change! Collective. See the introduction to the *Social Text* special issue on the topic: Marco Armiero et al., "Urban Climate Insurgency," *Social Text* 40 (2022).

22. For influential theoretical reflections on subaltern urban politics, see Asef Bayat, "Un-civil Society: The Politics of the 'Informal People,'" *Third World Quarterly* 18, no. 1 (1997): 53–72; and Bayat, "From 'Dangerous Classes' to 'Quiet Rebels': Politics of Urban Subalterns in the Global South," *International Sociology* 15, no. 3 (2000): 533–57.

23. Mike Davis, "Who Will Build the Ark?" *New Left Review* 61 (Jan/Feb 2010), https://newleftreview.org/issues/ii61/articles/mike-davis-who-will-build-the-ark.

24. Davis, *Planet of Slums*.

25. Sheela Patel and Sohanur Rahman, "The World's Poorest Have the Strongest Resilience, Yet Their Voices Remain Unheard," SDI, June, 6 2022, https://sdinet.org/2022/06/worlds-poorest-have-strongest-resilience-yet-voices-remain-unheard/.

26. Robinson, *Ordinary Cities*, x.

27. Edward Said, *Orientalism* (New York: Vintage, 1979).

28. Ananya Roy, *Poverty Capital: Microfinance and the Making of Development* (New York: Routledge, 2010); Colin MacFarlane, "The Entrepreneurial Slum: Civil Society, Mobility and the Co-production of Urban Development," *Urban Studies* 49, no. 13 (September 2012), https://doi.org/10.1177/0042098012452460,

29. Alexander Vasudevan, "The Makeshift City: Towards a Global Geography of Squatting," *Progress in Human Geography* 39, no. 3 (April 2014): 350.

30. Samuel Stein, *Capital City: Gentrification and the Real Estate State* (London: Verso Books, 2019), 2.

31. Saskia Sassen, *Expulsions: Brutality and Complexity in the Global Economy* (Cambridge, MA: Belknap Press, 2014); Matthew Desmond, *Evicted: Poverty and Profit in the American City* (New York: Crown Publishers, 2016); and Utku Balaban, "The Enclosure of Urban Space and Consolidation of the Capitalist Land Regime in Turkish Cities," *Urban Studies* 48, no. 10 (December 2010): 2162–79.

32. Liza Weinstein, "Evictions: Reconceptualizing Housing Insecurity from the Global South," *City & Community* 20, no. 1 (February 2021), 13–23.

33. Arjun Appadurai, "Spectral Housing and Urban Cleansing: Notes on Millenial Mumbai," *Public Culture* 12, no. 3 (2000): 627–51.

34. See, for example, Holston, *Insurgent Citizenship* and Vasudevan, "Makeshift City."

35. AbdouMaliq Simone, "People as Infrastructure: Intersecting Fragments in Johannesburg," *Public Culture* 16, no. 3 (2004): 407–29.

36. Jonathan R. Woetzel, *A Blueprint for Addressing the Global Affordable Housing Challenge* (New York: McKinsey Global Institute, 2014), 2.

37. Tom Gillespie, "Accumulation by Urban Dispossession: Struggles over Urban Space in Accra, Ghana," *Transactions of the Institute of British Geographers* 41, no. 1 (2016): 66–77. On accumulation by dispossession in general, see Louis Moreno and Hyun Bang Shin, "Introduction: The Urban Process under Planetary Accumulation by Dispossession," *City* 22,

no. 1 (2018): 78–87; and David Harvey, *The New Imperialism* (New York: Oxford University Press, 2003).

38.  Gillespie, "Accumulation by Urban Dispossession," 67.

39.  Sapana Doshi, "Greening Displacements, Displacing Green: Environmental Subjectivity, Slum Clearance, and the Embodied Political Ecologies of Dispossession in Mumbai," *International Journal of Urban and Regional Research* 42, no. 1 (January 2019): 112–32, https://doi.org/10.1111/1468-2427.12699.

40.  Davis, *Planet of Slums*, 19.

41.  Davis, *Planet of Slums*, 121.

42.  Davis, *Planet of Slums*, 134.

43.  Camilo Mora et al., "Global Risk of Deadly Heat," *Nature Climate Change* 7, no. 7 (2017): 501–6, https://doi.org/10.1038/nclimate3322.

44.  Robert Neuwirth, *Shadow Cities: A Billion Squatters, a New Urban World* (New York: Routledge, 2006).

45.  Davis, *Planet of Slums*, 139.

46.  Frantz Fanon, *The Wretched of the Earth* (New York: Grove Press, 1968), 38.

47.  World Health Organization, "Drinking-Water," https://www.who.int/news-room/fact-sheets/detail/drinking-water.

48.  Davis, *Planet of Slums*, 68.

49.  Alain R. A. Jacquemin, *Urban Development and New Towns in the Third World* (New York: Routledge, 1999), 41, 65.

50.  Davis, *Planet of Slums*, 137.

51.  Beth Gardiner, "The Deadly Cost of Dirty Air," *National Geographic* 239, no. 4 (April 2021): 48.

52.  Cited in Sean Sweeney and John Treat, *The Road Less Travelled: Reclaiming Public Transport for Climate Ready Mobility*, Trade Unions for Energy Democracy, May 2019, https://unionsforenergydemocracy.org/resources/tued-working-papers/tued-working-paper-12/, p. 10.

53.  Sweeney and Treat, *Road Less Travelled*, 10.

54.  Cited in Sweeney and Treat, *Road Less Travelled*, 11.

55.  Achille Mbembe, *Necropolitics* (Durham, NC: Duke University Press, 2019).

56.  Sweeney and Treat, *Road Less Travelled*, 25.

57.  Quoted in Sweeney and Treat, *Road Less Travelled*, 13.

58.  Nasser Abourahme, "Of Monsters and Boomerangs: Colonial Returns in the Late Liberal City," *City* 22, no. 1 (March 2018): 106–15, https://doi.org/10.1080/13604813.2018.1434296.

59.  For a historical overview of the demolition of Maroko, see Nathaniel Akhigbe, "How Nigerian Elites Grabbed Maroko Land 25 Years Ago," *Business Day*, August 16, 2015, https://businessday.ng/news/news-features/

article/how-nigerian-elites-grabbed-maroko-land-25-years-ago-part-1/.

60. Daniel Immerwahr, "The Politics of Architecture and Urbanism in Post-colonial Lagos, 1960–1986," *Journal of African Cultural Studies* 19, no. 2 (December 2007): 179.

61. Chris Abani, *Graceland* (New York: Picador, 2005).

62. Immerwahr, "Politics of Architecture and Urbanism," 177.

63. Akhigbe, "How Nigerian Elites Grabbed Maroko."

64. Akhigbe, "How Nigerian Elites Grabbed Maroko."

65. World Resources Institute, *World Resources Report: Towards a More Equitable City* (October 19, 2021), https://doi.org/10.46830/wrirpt.19.00124

66. For case studies of people's movements in cities of the Global South, see Trevor Ngwane, Luke Sinwell, and Immanuel Ness, eds., *Urban Revolt: State Power and the Rise of People's Movements in the Global South* (Chicago: Haymarket Books, 2017).

67. See Yves Cabannes, Silvia Guimarâes Yafai, and Cassidy Johnson, eds., *How People Face Evictions* (Development Planning Unit/University College London, 2010), https://www.ucl.ac.uk/bartlett/development/case-studies/2011/nov/how-people-face-evictions.

68. Cabannes, Guimarâes, and Johnson, *How People Face Evictions*, 3–4.

69. Cabannes, Guimarâes, and Johnson, *How People Face Evictions*, 32. For more on South African people's movements, see Trevor Ngwane, *Amakomiti: Grassroots Democracy in South African Shack Settlements* (New York: Pluto, 2021).

70. Cited in Robin King et al., *Confronting the Urban Housing Crisis in the Global South: Adequate, Secure, and Affordable Housing*, World Resources Institute, July 2017.

71. King et al., *Confronting the Urban Housing Crisis*.

72. Simone, "People as Infrastructure."

73. Examples of positive accounts of informal settlements' environmental characteristics are relatively sparse but include David Wachsmuth, Daniel Aldana Cohen, and Hillary Angelo, "Expand the Frontiers of Urban Sustainability," *Nature* 536 (2016): 391–93, https://doi.org/10.1038/536391a; and Tim Smedley, "Sustainable Urban Design: Lessons to Be Taken from Slums," *The Guardian*, June 5, 2013.

74. Christopher Barrington-Leigh and Adam Millard-Ball, "Global Trends toward Urban Street-Network Sprawl," *Proceedings of the National Academy of Sciences* 117, no. 4 (January 2020): 1941–50.

75. See, for example, Alex Smith, "China's Poorly Planned Cities: Urban Sprawl and the Rural Underclass Left Behind," *SupChina*, March 11, 2020.

76. Wachsmuth, Aldana Cohen, and Angelo, "Expand the Frontiers of Urban Sustainability."

77. J. F. C. Turner, "The Squatter Settlement," *Architectural Design* 38 (1968): 357.

78. Turner, "Squatter Settlement," 360.

79. Turner, "Squatter Settlement," 357.

80. Justin McGuirk, *Radical Cities across Latin America in Search of a New Architecture* (New York: Verso Books, 2014), 11.

81. Davis, *Planet of Slums*, 72.

82. McGuirk, *Radical Cities*, 25.

83. King et al., *Confronting the Urban Housing Crisis.*

84. Davis, "Who Will Build the Ark?"

85. City of Cape Town, "About Informal Housing."

86. Urban-Think Tank, "Empower Shack," Architizer, December 13, 2020, https://architizer.com/projects/empower-shack.

87. Urban-Think Tank, "Empower Shack."

88. Urban-Think Tank, "Empower Shack."

89. Urban-Think Tank, "Empower Shack."

90. Jose Collado and Han-Hsiang Wang, "Slum Upgrading and Climate Change Adaptation and Mitigation: Lessons from Latin America," *Cities* 104 (September 2020): https://doi.org/10.1016/j.cities.2020.102791, p. 10.

91. Slum Dwellers International, "Theory of Change," 2018, https://sdinet.org/publication/sdi-theory-change/.

92. King et al., *Confronting the Urban Housing Crisis*, 4.

93. McGuirk, *Radical Cities.*

94. Daniel Chavez, "Sweat Equity: How Uruguay's Housing Coops Provide Solidarity and Shelter to Low-Income Families," P2P Foundation, June 21, 2018, https://blog.p2pfoundation.net/sweat-equity-how-uruguays-housing-coops-provide-solidarity-and-shelter-to-low-income-families/2018/06/21.

95. Gustavo González, *Una Historia de FUCVAM* (Montevideo: Ediciones Trilce, 2013), 72.

96. González, *Una Historia de FUCVAM*, 137–64.

97. On urban politics during the Pink Tide, see Gianpaolo Baocchi, *We, the Sovereign* (Medford, MA: Polity Cress); and David Harvey, *Rebel Cities: From the Right to the City to the Urban Revolution* (London: Verso Books, 2012).

98. Collado and Wang, "Slum Upgrading," 6.

99. Collado and Wang, "Slum Upgrading," 6.

100. Collado and Wang, "Slum Upgrading," 8.

101. Collado and Wang, "Slum Upgrading," 8.

102. Collado and Wang, "Slum Upgrading," 8.

103. Smedley, "Sustainable Urban Design."

104. Collado and Wang, "Slum Upgrading," 8.

105. Abani, *Graceland*, 8–9.

106. Christo Venter, Anjali Mahendra, and Dario Hidalgo, "From Mobility to Access for All: Expanding Urban Transportation Choices in the Global South," World Resources Institute, May 14, 2019, 4.

107. Venter, Mahendra, and Hidalgo, "From Mobility to Access for All," 2.

108. Sweeney and Treat, *Road Less Travelled*, 12.

109. See UNICEF, *Silent Suffocation in Africa: Air Pollution Is a Growing Menace*, June 2019, https://www.unicef.org/reports/silent-suffocation-in-africa-air-pollution-2019.

110. Sweeney and Treat, *Road Less Travelled*, 10.

111. Venter, Mahendra, and Hidalgo, "From Mobility to Access for All," 4–5.

112. Venter, Mahendra, and Hidalgo, "From Mobility to Access for All," 4.

113. Venter, Mahendra, and Hidalgo, "From Mobility to Access for All," 5.

114. Drew Reed, "How Curitiba's BRT Stations Sparked a Transport Revolution – A History of Cities in 50 Buildings, Day 43," *The Guardian*, May 26, 2015.

115. Julie Gamble, "A Transit Manifesto for Quito," *NACLA Report on the Americas* 52, no. 2 (June 2020): 200.

116. Davis, "Who Will Build the Ark?" 30.

117. Davis, "Who Will Build the Ark?" 30.

118. Andreas Malm and the Zetkin Collective, *White Skin, Black Fuel: On the Danger of Fossil Fascism* (New York: Verso Books, 2021).

119. Brian Ashley et al., *One Million Climate Jobs: Moving South Africa Forward on a Low-Carbon, Wage-Led and Sustainable Path*, Alternative Information and Development Centre, December 2016.

120. Ashley et al., *One Million Climate Jobs*, 36.

121. Ashley et al., *One Million Climate Jobs*, 40.

122. Ashley et al., *One Million Climate Jobs*, 30.

123. Ashley et al., *One Million Climate Jobs*.

# Chapter 3: Reclaiming the Energy Commons

1. Gavin Bond, "Power to the People: Bernie Calls for Federal Takeover of Electricity Production," *Politico*, February 2, 2020.

2. Trevor Ngwane, personal interview, July 15, 2021.

3. For on overview of the ANC's shift to neoliberal governance, see Hein Marais, *South Africa Pushed to the Limit: The Political Economy of Change*

(Claremont, South Africa: UCT Press, 2011).

4.  In fact, Eskom launched its own Operation Khanyisa to encourage township residents to report their neighbors for electricity "theft." See "Operation Khanyisa to Smoke Out Izinyoka," *South African Government News*, October 26, 2010.

5.  Terri Maggott et al., *Energy Racism: The Electricity Crisis and the Working Class in South Africa*, Centre for Sociological Research and Practice, April 2022.

6.  Maggott et al., *Energy Racism*, 41.

7.  Maggott et al., *Energy Racism*, ii.

8.  Bond, "Power to the People."

9.  Quoted in Tracy Ledger, *Broken Promises*, Public Affairs Research Institute, April 2021, 9.

10. Vishwas Satgar, "Reclaiming the South African Dream," *Socialist Project*, January 2, 2012.

11. Shauna Mottiar, "Shaping a Township: Self-Connecting as 'Counter Conduct' in Umlazi, Durban," *Journal of the British Academy* 9, no. 11 (2021): 94.

12. For quotations drawn from township residents that express feelings of betrayal, see Maggott et al., *Energy Racism*.

13. On political fragmentation in post-apartheid South Africa, see Marcel Paret, "Resistance within South Africa's Passive Revolution: From Racial Inclusion to Fractured Militancy," *International Journal of Politics, Culture, and Society* 35, no. 4 (2022): 567–89, https://doi.org/10.1007/s10767-021-09410-x.

14. "Cruel optimism" is a term developed by the cultural critic Lauren Berlant to describe the continued investment of oppressed communities and people in the US in the Democratic Party. She did not offer a global gloss on the idea, but I believe it offers a useful lens on postcolonial history. See Lauren Berlant, *Cruel Optimism* (Durham, NC: Duke University Press, 2012).

15. On the corruption and devastating impact of the Narmada Valley Project, see Arundhati Roy, *Power Politics* (Boston, MA: South End Press, 2002).

16. Frantz Fanon, *Wretched of the Earth* (New York: Grove Press, 1963), 164–69, 181–83.

17. Colin Raymond, "The Emergence of Heat and Humidity Too Severe for Human Tolerance," *Science Advances* 6, no. 19 (May 2020): https://www.science.org/doi/full/10.1126/sciadv.aaw1838.

18. Eun-Soon Im, Jeremy S. Pal, and Elfatih A. B. Eltahir, "Deadly Heat Waves Projected in the Densely Populated Agricultural Regions of South Asia," *Science Advances* 3, no. 8 (August 2017): https://www.science.org/doi/full/10.1126/sciadv.1603322.

19. Dan Welsby et al., "Unextractable Fossil Fuels in a 1.5 °C World," *Nature* 597, no. 7875 (September 2021): 230–34, https://doi.org/10.1038/s41586-021-03821-8.

20. International Renewable Energy Association, *Tracking SDG 7: The Energy Progress Report* (Washington, DC: World Bank, 2022).

21. Brototi Roy and Anke Schaffartzik, "Talk Renewables, Walk Coal: The Paradox of India's Energy Transition," *Ecological Economics* 180 (February 2021): 1.

22. Andrea Cardoso and Ethemcan Turhan, "Examining New Geographies of Coal: Dissenting Energyscapes in Colombia and Turkey," *Applied Energy* 224 (August 2018): 398–408.

23. For one exemplary account of this history, see Timothy Mitchell, *Carbon Democracy: Political Power in the Age of Oil* (London: Verso Books, 2013).

24. Nicole Fabricant and Bret Gustafson, "Revolutionary Extractivism in Bolivia?" *NACLA*, March 2, 2015.

25. Naomi Klein, *This Changes Everything: Capitalism versus the Climate* (New York: Simon & Schuster, 2014).

26. Leah Temper et al., "Movements Shaping Climate Futures: A Systematic Mapping of Protests against Fossil Fuel and Low-Carbon Energy Projects," *Environmental Research Letters* 15, no. 12 (2020): 16.

27. For more extensive analysis of histories of global solidarity, see David Featherstone, *Solidarity: Hidden Histories and Geographies of Internationalism* (London: Zed Books, 2012).

28. Temper et al., "Movements Shaping Climate Futures," 1.

29. Temper et al., "Movements Shaping Climate Futures."

30. Mitchell, *Carbon Democracy*.

31. Alok Prakash Putul, "As India Faces Coal Shortages, a Mine Extension Has Been Approved in the Pristine Hasdeo Forests," *Scroll.in*, May 29, 2022.

32. Quoted in Putul, "As India Faces Coal Shortages."

33. Brian Cassey, "India's Ancient Tribes Battle to Save Their Forest Home from Mining," *The Guardian*, February 10, 2020.

34. Press Information Bureau Delhi, "Unleashing Coal: New Hopes for Atmanirbhar Bharat," press release, June 11, 2020, https://pib.gov.in/PressReleasePage.aspx?PRID=1630919."

35. Press Information Bureau Delhi, "Unleashing Coal."

36. Satya Sontanam, "All You Want to Know about Coal Mine Auctions," *Hindu BusinessLine*, June 22, 2020.

37. Welsby et al., "Unextractable Fossil Fuels."

38. Reuters, "UN Chief Urges Wealthy Nations to Phase Out Coal Use by 2030," *Mining Weekly*, March 3, 2021, https://www.miningweekly.

com/article/un-chief-urges-wealthy-nations-to-phase-out-coal-use-by-2030-2021-03-03/rep_id:3650.

39. Steven J. Davis et al., "Emissions Rebound from the COVID-19 Pandemic," *Nature Climate Change* 12, no. 5 (May 2022): 412–14, https://doi.org/10.1038/s41558-022-01332-6.

40. Roy and Schaffartzik, "Talk Renewables, Walk Coal."

41. Roy and Schaffartzik, "Talk Renewables, Walk Coal."

42. Cardoso and Turhan, "Examining New Geographies of Coal."

43. Al Jazeera Staff, "India Unveils Renewable Energy Ambitions," Al Jazeera, November 3, 2021.

44. Roy and Schaffartzik, "Talk Renewables, Walk Coal," 2.

45. Roy and Schaffartzik, "Talk Renewables, Walk Coal," 4.

46. Manjunath Bomnalli, "Coal India Will Not Be Privatized," *Deccan Herald*, July 10, 2020.

47. International Energy Agency, "India Has the Opportunity to Build a New Energy Future," press release, February 9, 2021.

48. Quoted in "India Has the Opportunity to Build a New Energy Future ."

49. Karl Mathiesen, "The Last-Minute Coal Demand That Almost Sunk the Glasgow Climate Deal," *Politico*, November 13, 2021.

50. Hannah Ellis-Petersen, "Indian Criticised Over Coal at COP26 – But the Real Villain Was Climate Injustice," *The Guardian*, November 14, 2021.

51. Joshua W, Busby et al., "The Case for US Cooperation with India on a Just Transition Away from Coal," Brookings, April 20, 2021.

52. Abinash Mohanty and Shreya Wadhawan, "Mapping India's Climate Vulnerability," *Hindustan Times*, November 11, 2021.

53. Nick Scott and Jordan Mendys, "This Is What It Means to Be Poor in India Today," CNN, October 2017.

54. Bhasker Tripathi, "India 5th Most Vulnerable to Climate Change Fallouts, Its Poor the Worst Hit," *India Spend*, December 5, 2019.

55. Abinash Mohanty, *Preparing India for Extreme Climate Events*," CEEW, December 2020.

56. Busby et al., "Case for US Cooperation."

57. Shankar Gopalakrishnan, "The Forest Rights Act," *Economic and Political Weekly* 52, no. 31 (June 2015): 7–8.

58. Gopalakrishnan, "Forest Rights Act."

59. Gopalakrishnan, "Forest Rights Act."

60. James C. Scott, *Seeing Like a State: How Certain Schemes to Improve the Human Condition Have Failed* (New Haven, CT: Yale University Press, 1999).

61. Gopalakrishnan, "Forest Rights Act."

62. Amnesty International, "When Land Is Lost, Do We Eat Coal? Coal

Mining and Violation of Adivasi Rights in India," July 11, 2016.

63. Matthew Shutzer and Arpitha Kodiveri, "A Vast Bed of Combustible Fuel," *Radical History Review* 145 (January 2023): 13–36.

64. Shalini Gera, personal interview.

65. Amnesty International, "When Land Is Lost."

66. Jo Woodman, "In Modi's India, Being a Tribal Woman Is an Act of Resistance," *CounterPunch*, July 9, 2021.

67. Abhinav Sekhri, "How the UAPA Is Perverting the Idea of Justice," *Article14*, July 16, 2022, https://www.article-14.com/post/how-the-uapa-is-perverting-india-s-justice-system.

68. Sumedha Pal, "One Year after Arrest, Organisations Demand Adivasi Activist Hidme Markam's Release" *The Wire*, March 9, 2022.

69. Asian Centre for Human Rights, *The State of Encounter Killings in India: Target, Detain, Torture, Execute*. See also Amnesty International, "India Report," European Country of Information Network, November 1, 2019, https://www.ecoi.net/en/document/1457651.html.

70. Alpa Shah, "The Agrarian Question in a Maoist Guerrilla Zone: Land, Labour and Capital in the Forests and Hills of Jharkhand, India," *Journal of Agrarian Change* 13 (June 2013): 424–50, https://doi.org/10.1111/joac.12027.

71. Shah, "Agrarian Question," 447.

72. Nandini Sundar, *The Burning Forest: India's War against the Maoists* (London: Verso Books, 2019), xii.

73. Andrea Pitzer, "Concentration Camps Existed Long Before Auschwitz," *Smithsonian* (magazine), November 2, 2017.

74. See Sundar, *Burning Forest*, and Arundhati Roy, *Walking with the Comrades* (New York: Penguin Books, 2011).

75. Harsh Mander, "Urban Maoists," *South China Morning Post*, August 31, 2018.

76. EJOLT, "Coal Mining Conflict in Hazaribagh with NTPC in Jharkhand, India," Environmental Justice Atlas, October 9, 2016.

77. Arnim Scheidel et al., "Environmental Conflicts and Defenders: A Global Overview," *Global Environmental Change* 63 (July 2020), https://www.sciencedirect.com/science/article/pii/S0959378020301424.

78. Brototi Roy and Joan Martinez-Alier, "Environmental Justice Movements in India: A Systematic Mapping of Protests against Fossil Fuel and Low-Carbon Energy Projects," *Ecology, Economy and Society – the INSEE Journal* 2, no. 1 (January 2019): 77–92, https://doi.org/10.37773/ees.v2i1.56.

79. Roy and Martinez-Alier, "Environmental Justice Movements in India."

80.  Scheidel et al., "Environmental Conflicts and Defenders."

81.  Scheidel et al., "Environmental Conflicts and Defenders."

82.  Brototi Roy, personal interview.

83.  StopAdani, "Timeline of #StopAdani Actions," https://www.tiki-toki.com/timeline/entry/1006867/Timeline-of-StopAdani-actions.

84.  Shaun Griswold, "Interior Department Report Details the Brutality of Federal Indian Boarding Schools," *Wisconsin Examiner*, May 17, 2022.

85.  Roy and Schaffartzik, "Talk Renewables, Walk Coal," 6.

86.  Roy and Schaffartzik, "Talk Renewables, Walk Coal," 7.

87.  David Pegg, "Why the Mundra Power Plant Has Given Tata a Mega Headache," *The Guardian*, April 16, 2015.

88.  Barbara Grzincic, "Indian Coal Plant's World Bank Lender Immune from Enviro Suit," Reuters, July 7, 2021.

89.  Patrik Oskarsson et al., "India's New Coal Geography: Coastal Transformations, Imported Fuel and State-Business Collaboration in the Transition to More Fossil Fuel Energy," *Energy Research & Social Science* 73 (March 2021): 1.

90.  Karishma Mehrotra, "In India's Coal Belts, Jobs Are Now Hard to Get—and Harder to Keep," *Scroll.in*, February 10, 2022.

91.  Mehrotra, "In India's Coal Belts."

92.  Martín Arboleda, *Planetary Mine: Territories of Extraction under Late Capitalism* (London: Verso Books, 2020), 3–4.

93.  Rajini Vaidyanathan, "Climate Change: Why India Can't Live without Coal," BBC News, September 28, 2021, https://www.bbc.com/news/world-asia-india-58706229.

94.  Just Transition Initiative Team, "Understanding Just Transitions in Coal Dependent Communities," October 2021, https://justtransitioninitiative.org/understanding-just-transitions-in-coal-dependent-communities/.

95.  Roy and Schaffartzik, "Talk Renewables, Walk Coal."

96.  "How a Just Transition Can Make India's Coal History," BBC, November 9, 2021.

97.  Quoted in Patrick Greenfield, "South African Environmental Activist Shot Dead," *The Guardian*, October 23, 2020.

98.  Global Witness, *Last Line of Defence*, (London: Global Witness), 2021.

99.  For an analysis of mining and the oppression of women, see WoMin, *Extractives and Violence against Women*, March 2022.

100. Mary Lawlor, *Final Warning: Death Threats and Killings of Human Rights Defenders: Report of the Special Rapporteur on the Situation of Human Rights Defenders* (Geneva: United Nations, 2020).

101. Lawlor, *Final Warning*, 3.

102. Katharina Rall and Ramin Pejan, *We Know Our Lives Are in Danger: Environment of Fear in South Africa's Mining-Affected Communities* (New York: Human Rights Watch, 2019).

103. Rall and Pejan, *We Know Our Lives Are in Danger*.

104. Centre for Environmental Rights, "A Momentous Legal Victory for Environmental Activism and Free Speech," February 10, 2021.

105. Rob Nixon, *Slow Violence and the Environmentalism of the Poor* (Cambridge, MA: Harvard University Press, 2013).

106. Save our Wilderness, "ZAC's Christmas Contamination Crisis," January 4, 2022, https://saveourwilderness.org/2021/12/31/zacs-christmas-contamination-crisis/.

107. WoMin, *No Longer a Life Worth Living* (Johannesburg, SA: WoMin African Gender and Extractives Alliance, 2016).

108. WoMin, *No Longer a Life Worth Living*, vi.

109. Sam Ashman, "SA's Climate Crisis Is Embedded in Coal and Exports," *New Frame*, August 30, 2021.

110. Ben Fine and Zavareh Rustomjee, *The Political Economy of South Africa: From Minerals-Energy Complex to Industrialisation* (New York: Routledge, 1996).

111. groundWork, *Coal Kills: Research and Dialogue for a Just Transition* (Pietermaritzburg, SA: groundWork, 2018).

112. groundWork, *Coal Kills*, 5.

113. Caryle Murphy, "To Cope with Embargoes, S. Africa Converts Coal into Oil," *Washington Post*, April 27, 1979.

114. Ashley van Niekerk et al., "Energy Impoverishment and Burns: The Case for an Expedited, Safe and Inclusive Energy Transition in South Africa," *South African Journal of Science* 118, no. 3–4 (March 2022), https://doi.org/10.17159/sajs.2022/13148.

115. World Resources Institute, "South Africa: Strong Foundations of a Just Transition," December 23, 2021.

116. Ashman, "SA's Climate Crisis"; and Jean Imbs, "The Premature Deindustrialization of South Africa," in *The Industrial Policy Revolution II: Africa in the Twenty-first Century*, ed. Joseph E. Stiglitz, Justin Yifu Lin, and Ebrahim Patel, International Economic Association Series (London: Palgrave Macmillan, 2013).

117. Charles Dednam, "COVID-19: South African Steel Industry," Trade and Industrial Policy Strategies (July 2020).

118. Dednam, "South African Steel Industry."

119. Seeraj Mohamed, "The Financialization of the South African Economy," *Development* 59, no. 1 (April 2016): 39.

120. Sam Ashman, Ben Fine, and Susan Newman, "The Crisis in South Africa: Neoliberalism, Financialization and Uneven and Combined Development" *Socialist Register* 47 (2011): 177.
121. Ashman, Fine, and Newman, "Crisis in South Africa," 177.
122. Ed Stoddard, "Minerals and Energy: The Decline of South Africa's Mining Sector in Five Charts," *Daily Maverick*, November 23, 2021.
123. Centre for Development and Enterprise, "VIEWPOINTS | Reviving a Declining Mining Industry," *CDE – The Centre for Development and Enterprise* (blog), August 4, 2020.
124. Prinesha Naidoo and Felix Njini, "Iconic South African Mines Ravaged Economy's Unlikely Savior," *Bloomberg*, July 6, 2021, https://www.bloomberg.com/news/articles/2021-07-06/iconic-south-african-mines-are-ravaged-economy-s-unlikely-savior.
125. Centre for Development and Enterprise, "VIEWPOINTS."
126. Jeremy Baskin, *Striking Back: A History of Cosatu* (New York: Verso Books, 1991). 224–39.
127. Alternative Information and Development Centre, "Migrant Marikana," August 23, 2017, AIDChttps://aidc.org.za/migrant-marikana-shifts-migrant-labour-system/.
128. Arboleda, *Planetary Mine*, 2–3.
129. Stoddard, "Minerals and Energy."
130. Paret, "Resistance within South Africa's Passive Revolution."
131. Van Niekerk, personal interview.
132. Alternative Information and Development Centre, "Their Just Transition and Ours," AIDC, December 9, 2019, https://aidc.org.za/their-just-transition-and-ours/.
133. COSATU Central Executive Committee, "Cosatu: Congress of South African Trade Unions Policy Framework on Climate Change (19/11/2011)," November 19, 2011.
134. World Resources Institute, "South Africa."
135. Alternative Information and Development Centre, "Their Just Transition and Ours."
136. NUMSA, "Motivations for a Socially-Owned Renewable Energy Sector," October 15, 2022, https://numsa.org.za/2012/10/motivations-for-a-socially-owned-renewable-energy-sector-2012-10-15/.
137. Casey Williams, "Amid Rolling Blackouts, Energy Workers Fight for Clean Public Power in South Africa," *In These Times*, March 31, 2022.
138. Brian Ashley et al., *One Million Climate Jobs: Moving South Africa Forward on a Low-Carbon, Wage-Led and Sustainable Path*, Alternative Information and Development Centre, December 2016.

139. See, for instance, Kate Aronoff et al., *A Planet to Win: Why We Need a Green New Deal* (New York: Verso Books, 2019); Ann Pettifor, *The Case for the Green New Deal* (New York: Verso Books, 2019); and Mathew Lawrence and Laurie Laybourn-Langton, *Planet on Fire* (New York: Verso Books, 2022).

140. Mark Swilling et al., "Linking the Energy Transition and Economic Development: A Framework for Analysis of Energy Transitions in the Global South," *Energy Research & Social Science* 90 (August 2022), https://doi.org/10.1016/j.erss.2022.102567.

141. Ashley et al., *One Million Climate Jobs*, 26.

142. Ashley et al., *One Million Climate Jobs*, 26

143. David Roberts, "The Key to Tackling Climate Change," *Vox*, September 19, 2016.

144. Ashley et al., *One Million Climate Jobs*, 27–28.

145. Trade Unions for Energy Democracy, "TUED Working Paper #12," May 2019, 10.

146. Trade Unions for Energy Democracy, "TUED Working Paper #12," 51.

147. Mimi Sheller, *Mobility Justice: The Politics of Movement in an Age of Extremes* (New York: Verso Books, 2018), 83–87.

148. Ashley et al., *One Million Climate Jobs*, 41.

149. Ashley et al., *One Million Climate Jobs*, 47.

150. Ashley et al., *One Million Climate Jobs*, 49–51.

151. Ashley et al., *One Million Climate Jobs*, 56.

152. Ashley et al., *One Million Climate Jobs*, 56.

153. Van Niekerk, personal interview.

154. Ashley Dawson, "Opinion | Cape Town Has a New Apartheid," *Washington Post*, July 10 2018.

155. Van Niekerk, personal interview.

156. Karl Cloete, "Op-Ed: NUMSA Supports a Transition from Dirty Energy to Clean Renewable Energy," *Daily Maverick*, March 15, 2018.

157. Cloete, "NUMSA Supports."

158. Williams, "Amid Rolling Blackouts."

159. Maggott et al., *Energy Racism*, 48.

160. Maggott et al., *Energy Racism*, ii.

161. BizNews, "How to Steal a Billion from Eskom—and Leave SA in Darkness," *BizNews*, April 20, 2022, https://www.biznews.com/energy/2022/04/20/losing-power-steal-eskom-darkness.

162. Eskom Research Reference Group, *Eskom Transformed: Achieving a Just Transition for South Africa*, July 2020, 15.

163. Brian Kamanzi, "Collapse of Energy Utilities Sounds a Warning Bell,"

*New Frame*, February 18, 2022, https://www.newframe.com/collapse-of-energy-utilities-sounds-a-warning-bell.

164. Oladimeji Joseph Ayamolowo, P.T. Manditereza, and K. Kusakana, "South Africa Power Reforms: The Path to a Dominant Renewable Energy-Sourced Grid," *Energy Reports* 8 (April 2022): 1208–15, https://doi.org/10.1016/j.egyr.2021.11.100.

165. Rolando Fuentes-Bracamontes, "Is Unbundling Electricity Services the Way Forward for the Power Sector?" *Electricity Journal* 29, no. 9 (November 2016): 16–20, https://doi.org/10.1016/j.tej.2016.10.006.

166. Kamanzi, "Collapse of Energy Utilities Sounds a Warning Bell."

167. Naomi Klein, "Why Texas Republicans Fear the Green New Deal," *New York Times*, February 21, 2021.

168. Bruce Baigrie, "Eskom, Unbundling, and Decarbonization," *Phenomenal World*, February 14, 2022, https://www.phenomenalworld.org/analysis/eskom-unbundling-and-decarbonization/.

169. COP26, "Political Declaration on the Just Energy Transition in South Africa," UN Climate Change Conference (COP26) at the SEC, November 2, 2021, https://ukcop26.org/political-declaration-on-the-just-energy-transition-in-south-africa/.

170. Craig Morris and Arne Jungjohann, *Energy Democracy: Germany's Energiewende to Renewables* (New York: Springer International, 2018).

171. Eskom Research Reference Group, *Eskom Transformed*, 39.

172. Eskom Research Reference Group, *Eskom Transformed*, 58.

173. Eskom Research Reference Group, *Eskom Transformed*, 62.

174. See, for example, Victoria Masterson, "'Renewables' Power Ahead to Become the World's Cheapest Source of Energy in 2020," World Economic Forum, July 5, 2021, https://www.weforum.org/agenda/2021/07/renewables-cheapest-energy-source/.

175. Eskom Research Reference Group, *Eskom Transformed*, 49.

176. Veronika Henze, "Global Investment in Low-Carbon Energy Transition Hit $755 Billion in 2021," *Bloomberg*, January 27, 2022.

177. Staff, "Global Landscape of Climate Finance 2019," Climate Policy Initiative, November 7, 2019, https://www.climatepolicyinitiative.org/publication/global-landscape-of-climate-finance-2019/.

178. Eskom Research Reference Group, *Eskom Transformed*, 83–129.

179. Sweeney, personal interview.

180. Ngwane, personal interview.

181. International Renewable Energy Agency, *RE-organising Power Systems for the Transition*, June 2022, 15, https://www.irena.org/publications/2022/Jun/RE-organising-Power-Systems-for-the-Transition.

182. International Renewable Energy Agency, *RE-organising Power Systems for the Transition*, 16.

183. Maggott et al., *Energy Racism*, 32.

184. Maggott et al., *Energy Racism*, ii.

185. Ruppert Bulmer et al., *Global Perspective on Coal Jobs and Managing Labor Transition out of Coal*, World Bank, Washington, DC, December 2021.

186. Ruth Maclean and Dionne Searcey, "Congo to Auction Land to Oil Companies: Our Priority Is Not to Save The Planet," *New York Times*, July 24, 2022.

187. Maclean and Searcey, "Congo to Auction Land."

188. David Hill, "Ecuador Pursued China Oil Deal While Pledging to Protect Yasuni, Papers Show," *The Guardian*, February 19, 2014.

189. Greg Muttitt and Sivan Kartha, "Equity, Climate Justice and Fossil Fuel Extraction: Principles for a Managed Phase Out," *Climate Policy* 20, no. 8 (September 2020): 1024–42, https://doi.org/10.1080/14693062.2020.1 763900.

190. Oxfam, "Poorer Nations Expected to Face up to $75 Billion Six-Year Shortfall in Climate Finance," September 19, 2021, https://www.oxfama-merica.org/press/poorer-nations-expected-to-face-up-to-75-billion-six-year-shortfall-in-climate-finance-oxfam/.

191. Bulmer et al., *Global Perspective on Coal Jobs*, 2.

192. Bulmer et al., *Global Perspective on Coal Jobs*, 12.

193. Keith Schneider, "World Bank, Despite Promises, Finances Big Coal and Industrial Projects That Threaten Water, Communities," *Circle of Blue*, October 18, 2016, https://www.circleofblue.org/2016/world/world-bank-despite-promises-finances-big-coal-industrial-projects-threat-en-water-communities.

194. Luisa Abbott Galvao, "The Hidden Flows of Finance to Fossil Fuels: World Bank and IMF Edition," Friends of the Earth, May 25, 2021, https://foe.org/blog/the-hidden-flows-of-finance-to-fossil-fuels-world-bank-and-imf-edition/.

195. Swaraj Singh Dhanjal, "Adani Group Raises $9 Billion from Offshore Bond Market," *mint*, October 6, 2021, https://www.livemint.com/companies/news/adani-group-raises-9-billion-from-offshore-bond-mar-ket-11633457139162.html.

196. Bhumika Muchhala, "Towards a Decolonial and Feminist Global Green New Deal," Rosa-Luxemburg-Stiftung, August 24, 2020, https://www.rosalux.de/en/news/id/43146/towards-a-decolonial-and-feminist-global-green-new-deal.

## Chapter 4: Against Fortress Conservation

1.  Justin Rowlatt, "Kaziranga: The Park That Shoots People to Protect Rhinos," BBC, February 10, 2017, https://www.bbc.com/news/world-south-asia-38909512.
2.  Arunabh Saikia, "Kaziranga Activists Jailed: Colleagues Claim This Is Vendetta for Their Role in BBC Film on Poaching," *Scroll.in*, May 5, 2017, https://scroll.in/article/836565/kaziranga-activists-jailed-were-they-held-for-featuring-in-bbc-film-critical-of-poaching-policy.
3.  Saikia, "Kaziranga Activists Jailed."
4.  NET Web Desk, "Assam: Commandos Will Be Deployed in Kaziranga National Park to Combat Rhino Poaching," *Northeast Today*, January 22, 2022, https://www.northeasttoday.in/2022/01/24/assam-commandos-will-be-deployed-in-kaziranga-national-park-to-combat-rhino-poaching/.
5.  Anupam Chakravartty, "Latest Kaziranga Expansion Brings Back Fears of Evictions among Residents," *Mongabay*, November 25, 2020, https://india.mongabay.com/2020/11/latest-kaziranga-expansion-brings-back-fear-of-evictions-among-residents/.
6.  Sushanta Talukdar, "Waiting for Curzon's Kin to Celebrate Kaziranga," *The Hindu*, January 5. 2005.
7.  Kaziranga National Park, "History of Kaziranga National Park," https://www.kaziranga-national-park.com/kaziranga-history.shtml.
8.  Joëlle Smadja, "A Chronicle of Law Implementation in Environmental Conflicts: The Case of Kaziranga National Park in Assam (North-East India)," *South Asia Multidisciplinary Academic Journal* 17 (2018): 1–37.
9.  Chakravartty, "Latest Kaziranga Expansion."
10. BBC Newsnight, "How Far Should We Go to Stop Poaching?" YouTube video, February 10, 2017, https://www.youtube.com/watch?v=2EdfVvK1Ft0.
11. Rudolfo Dirzo et al., "Defaunation in the Anthropocene," *Science* 345, no. 6195 (July 2014): 401–6.
12. R E A Almond, M. Grooten, and T. Petersen, eds. *Living Planet Report 2020: Bending the Curve of Biodiversity Loss* (Gland, Switzerland: WWF, 2020), https://www.worldwildlife.org/publications/living-planet-report-2020.
13. Dirzo et al., "Defaunation in the Anthropocene," 401.
14. Royal Botanic Gardens, Kew, "State of the World's Plants and Fungi 2020," https://www.kew.org/science/state-of-the-worlds-plants-and-fungi.
15. Dirzo et al., "Defaunation in the Anthropocene," 401.
16. James W. Moore, *Anthropocene or Capitalocene?: Nature, History, and the Crisis of Capitalism* (Oakland, CA: PM Press/Kairos, 2016).

17. Mark Dowie, "Conservation Refugees," *Orion Magazine*, February 21, 2015, https://orionmagazine.org/article/conservation-refugees/.

18. Protected Planet, "Executive Summary," in *Protected Planet Report 2020*, May 2021, https://livereport.protectedplanet.net/chapter-1.

19. Eric Dinerstein et al., "An Ecoregion-Based Approach to Protecting Half the Terrestrial Realm," *Bioscience* 67, no. 6 (June 2017): 534–45.

20. Dinerstein et al., "Ecoregion-Based Approach," 534.

21. Survival International, *Our Land, Our Nature! A People's Manifesto for the Future of Conservation*, https://assets.survivalinternational.org/documents/2019/211013-olon-manifesto-en-es-fr.pdf.

22. Survival International, *Our Land, Our Nature!*

23. World Rainforest Management, "The Conservation Industry's Agenda in Times of Crisis," *WRM Bulletin* 249 (May 14, 2020), https://www.wrm.org.uy/bulletin-articles/the-conservation-industrys-agenda-in-times-of-crisis.

24. Richard Grove, *Green Imperialism: Colonial Expansion, Tropical Island Edens, and the Origins of Environmentalism, 1600–1860* (New York: Cambridge University Press, 1996).

25. Sidney Mintz, *Sweetness and Power: The Place of Sugar in Modern History* (New York: Penguin, 1986).

26. Gregg Mittman, Donna Haraway, and Anna Tsing, "Reflections on the Plantationocene: A Conversation with Donna Haraway and Anna Tsing," October 12, 2019, https://edgeeffects.net/wp-content/uploads/2019/06/PlantationoceneReflections_Haraway_Tsing.pdf.

27. Laws governing human conduct toward forests were in existence far before this, but never motivated by the kind of perception of environmental crisis that animated the framers of the Kings Hill Forest Act. On ancient Irish legal structures relating to trees, for example, see Tina R. Fields, "Trees in Early Irish Law and Lore: Respect for Other-Than-Human Life in Europe's History," *Ecopsychology* 12, no. 2 (2020): 130–37.

28. Quoted in Richard Grove, "The Culture of Islands and the History of Environmental Concern," in *Climate Change and the Humanities: Historical, Philosophical and Interdisciplinary Approaches to the Contemporary Environmental Crisis*, ed. Alexander Elliott, James Cullis, and Vinita Damodaran (New York: Palgrave Macmillan, 2017), 73.

29. Gregory Barton, *Empire Forestry and the Origins of Environmentalism* (New York: Cambridge University Press, 2002), 44.

30. Barton, *Empire Forestry*, 44.

31. B. H. Baden Powell, *Forest Law* (London: Bradbury Agnaw and Co., 1893), 225.

32. Barton, *Empire Forestry*, 45.
33. Barton, *Empire Forestry*, 46.
34. Barton, *Empire Forestry*, 65.
35. Quoted in E. P. Stebbing, *The Forests of India*, vol. 2 (London: John Lane,1923), 15.
36. Stebbing, *The Forests of India*, 16.
37. Peter Linebaugh, "Karl Marx, the Theft of Wood and Working-Class Composition: A Contribution to the Current Debate," *Crime and Social Justice* 6 (Fall-Winter 1979): 1–29.
38. Linebaugh, "Karl Marx, the Theft of Wood."
39. Linebaugh, "Karl Marx, the Theft of Wood," 12.
40. Linebaugh, "Karl Marx, the Theft of Wood," 14.
41. Cited in Judith Whitehead, "John Locke, Accumulation by Dispossession and the Governance of Colonial India," *Journal of Contemporary Asia* 42, no. 1 (February 2012): 10–11.
42. Whitehead, "John Locke," 10.
43. John Locke, *Second Treatise of Government*, ed. Thomas P. Peardon (New York: Liberal Arts Press, 1952), 22.
44. Barbara Arneil, *John Locke and America: The Defense of English Colonialism* (New York: Oxford University Press, 1998).
45. Locke, *Second Treatise of Government*, 22.
46. See Whitehead, "John Locke."
47. Vinay Gidwani, *Capital, Interrupted: Agrarian Development and the Politics of Work in India* (Minneapolis: University of Minnesota Press, 2008).
48. Barton, *Empire Forestry*, 40.
49. See Ranajit Guha, *A Rule of Property for Bengal: An Essay on the Idea of Permanent Settlement* (Durham, NC: Duke University Press, 1996).
50. Madhav Gadgil and Ramachandra Guha, *This Fissured Land: An Ecological History of India* (New York: Oxford University Press, 2012), 192.
51. Whitehead, "John Locke," 11.
52. Prakash Kashwan et al., "From Racialized Neocolonial Global Conservation to an Inclusive and Regenerative Conservation," *Environmental: Science and Policy for Sustainable Development* 63, no. 4 (2021): 6.
53. Ramachandra Guha offers a critical history that challenges the notion that environmental conservation began in the United States in *How Much Should a Person Consume? Environmentalism in India and the United States* (Berkeley: University of California Press, 2006).
54. Philip Burnham, *Indian Country, God's Country: Native Americans and the National Parks* (Washington, DC: Island Press, 2000).
55. Gadgil and Guha, *This Fissured Land*, 131.

56. Gadgil and Guha, *This Fissured Land*, 125.

57. Edward Buxton, *Two African Trips, with Notes and Suggestions on Big Game Preservation in Africa* (London: E. Stanford, 1902), 117.

58. Buxton, *Two African Trips*, 118.

59. Roderick P. Neumann, "Dukes, Earls, and Ersatz Edens: Aristocratic Nature Preservationists in Colonial Africa," *Environment and Planning D: Society and Space* 14, no. 1 (1996): 79–98.

60. Buxton, *Two African Trips*, 139.

61. Buxton, *Two African Trips*, 139.

62. Gadgil and Guha, *This Fissured Land*, 205.

63. Gadgil and Guha, *This Fissured Land*, 207.

64. Britta Sjostedt, "Rights of Indigenous Peoples and Environmental Protection in *Jus Post Bellum*," in *Just Peace after Conflict: Jus Post Bellum and the Justice of Peace*, ed. Carsten Stahn and Jens Iverson (Oxford: Oxford University Press, 2021).

65. Jenny Springer and Fernanda Almeida, *Protected Areas and the Land Rights of Indigenous Peoples and Local Communities: Current Issues and Future Agenda*, Rights and Resources Initiative, May 2015, https://rightsandresources.org/wp-content/uploads/RRIReport_Protected-Areas-and-Land-Rights_web.pdf.

66. Quoted in Sjostedt, "Rights of Indigenous Peoples."

67. Springer and Almeida, *Protected Areas and the Land Rights of Indigenous Peoples and Local Communities*, 3.

68. Mark Dowie, *Conservation Refugees: The Hundred-Year Conflict between Global Conservation and Native Peoples* (Cambridge, MA: MIT Press, 2009).

69. Tembi M. Tichaawa and Oswald Mhlanga, "Community Perceptions of a Community-Based Tourism Project: A Case Study of the CAMPFIRE Programme in Zimbabwe," *African Journal for Physical, Health Education, Recreation and Dance* 21, Supplement 1 (2015): 55.

70. Springer and Almeida, *Protected Areas and the Land Rights of Indigenous Peoples and Local Communities*, 15.

71. Springer and Almeida, *Protected Areas and the Land Rights of Indigenous Peoples and Local Communities*, 18.

72. Bram Büscher and Robert Fletcher, *The Conservation Revolution: Radical Ideas for Saving Nature beyond the Anthropocene* (New York: Verso Books, 2020), 20.

73. Büscher and Fletcher, *Conservation Revolution*, 20.

74. Büscher and Fletcher, *Conservation Revolution*, 21.

75. Büscher and Fletcher, *Conservation Revolution*, 21.

76. International Union for Conservation of Nature, *The Marseille Manifesto*, September 2021, https://www.iucncongress2020.org/programme/marseille-manifesto.

77. IUCN, *Marseille Manifesto*, 2.

78. IUCN, *Marseille Manifesto*, 3.

79. IUCN, *Marseille Manifesto*, 4.

80. IUCN, *Marseille Manifesto*, 4.

81. Judith Schleicher et al., "Protecting Half the Planet Could Directly Affect over One Billion People," *Nature Sustainability* 2 (December 2019): 1094–96.

82. Rainforest Foundation UK, "The 'Post-2020 Global Biodiversity Framework' – How the CBD Drive to Protect 30 Percent of the Planet Could Dispossess Millions," Mapping for Rights, July 2020, https://www.mappingforrights.org/MFR-resources/mapstory/cbddrive/300_million_at_risk_from_cbd_drive.

83. Schleicher et al., "Protecting Half the Planet," 1095.

84. Schleicher et al., "Protecting Half the Planet," 1096.

85. World Rainforest Management, "Conservation Industry's Agenda."

86. For a critical take on carbon offsetting that reflects many of the perspectives circulating in movements at the time, see Global Justice Ecology Project's 2012 film *A Darker Shade of Green: REDD Alert and the Future of Forests*, https://www.youtube.com/watch?v=FPFPUhsWMaQ.

87. Fiore Longo, "Why Nature-Based Solutions Won't Solve the Climate Crisis—They'll Just Make Rich People Richer," *Common Dreams*, October 13, 2021, https://www.commondreams.org/views/2021/10/13/why-nature-based-solutions-wont-solve-climate-crisis-theyll-just-make-rich-people.

88. Ashley Dawson, "Climate Justice: The Emerging Movement against Green Capitalism," *South Atlantic Quarterly* 109, no. 2 (April 2010): 313–38, https://doi.org/10.1215/00382876-2009-036.

89. IUCN, "Nature-based Solutions," https://www.iucn.org/theme/nature-based-solutions/about.

90. Jason Hickel, "The Contradiction of the Sustainable Development Goals: Growth Versus Ecology on a Finite Planet," *Sustainable Development* 27, no. 5 (September/October 2019): 873–84.

91. David Harvey, *The New Imperialism* (New York: Oxford University Press, 2003), 87.

92. United Nations, "Sustainable Development Goals," https://sdgs.un.org/goals/goal12.

93. See Shell Oil, "Nature-Based Solutions," https://www.shell.com/energy-and-innovation/new-energies/nature-based-solutions.html - if-

rame=L3dlYmFwcHMvMjAxOV9uYXR1cmVfYmFzZWRfc29sdXRp-
b25zL3VwZGF0ZS8.

94. Tom Dillon, "Glasgow Deal to Tackle Emissions Includes Nature-Based Solutions," Pew Charitable Trusts, November 18, 2021, https://www.pewtrusts.org/en/research-and-analysis/articles/2021/11/18/glasgow-deal-to-tackle-emissions-includes-nature-based-solutions.

95. Longo, "Why Nature-Based Solutions."

96. Longo, "Why Nature-Based Solutions."

97. Simon Counsell, "Anatomy of a 'Nature-Based Solution': Total Oil, 40,000 Hectares of Disappearing Savannah, Emmanuel Macron, Norwegian and French 'Aid' to an Election-Rigging Dictator, Trees to Burn, Secret Contracts, and Dumbstruck Conservationists," *REDD Monitor*, April 16, 2021, https://redd-monitor.org/2021/04/16/anatomy-of-a-nature-based-solution-total-oil-40000-hectares-of-disappearing-african-savannah-emmanuel-macron-norwegian-and-french-aid-to-an-election-rigging-dictator-trees/.

98. Quoted in Counsell, "Anatomy of a 'Nature-Based Solution.'" For TotalEnergies's own account of its NbS policies, see "Developing Activities That Contribute to Society's Carbon Neutrality," Future Net Zero, May 24, 2022, https://totalenergies.com/group/commitment/climate-change/carbon-neutrality.

99. Quoted in Counsell, "Anatomy of a 'Nature-Based Solution.'"

100. Wildlife Conservation Society, "Bateke Plateaux Landscape," https://congo.wcs.org/wild-places/bateke-plateaux.aspx.

101. Lauren Berlant, *Cruel Optimism* (Durham, NC: Duke University Press, 2012).

102. Gadgil and Guha, *This Fissured Land*, 130.

103. Gadgil and Guha, *This Fissured Land*, 135.

104. Gadgil and Guha, *This Fissured Land*, 136.

105. Oxfam, International Land Coalition, and Rights and Resources Initiative, *Common Ground: Securing Land Rights and Safeguarding the Earth* (Oxford: Oxfam, March 2016), https://rightsandresources.org/wp-content/uploads/2016/04/Global-Call-to-Action_Common-Ground_Land-Rights_April-2-16_English.pdf.

106. National Sample Survey Office, *India: Common Property Resources, Sanitation, and Hygiene Services, NSS 54th Round: January–June 1998*, Government of India, March 16, 2016, mospi.nic.in/rept%20_%20pubn/452_final.pdf.

107. Kumar Sambhav Shrivastava, "Govt to Allow PVT to Manage 40% of Forests," *Hindustan Times*, September 13, 2015, https://www.hindustan-

times.com/india/govt-toallow-pvt-sector-to-manage-40-of-forests/story-yOiG4TO4kA2kvykxXNTEBK.html.

108. Oxfam, *Common Ground*, 27.

109. International Land Coalition, *Tirana Declaration*, Global Assembly 2011, https://www.landcoalition.org/en/about-ilc/governance/assemblydeclarations/2011-tirana/.

110. Oxfam, *Common Ground*, 27.

111. Nia Emmanouil and Carla Chan Unger, *First Peoples and Land Justice Issues in Australia: Addressing Deficits in Corporate Accountability*, RMIT University, March 2021, https://www.rmit.edu.au/research/centres-collaborations/business-and-human-rights-centre/research-projects/first-peoples-land-justice-issues-australia.

112. Rights and Resources Initiative, "Who Owns the Land in Africa?" RRI Factsheet, October 2015, https://rightsandresources.org/publication/who-owns-the-land-in-africa/.

113. Survival International, *Our Land, Our Nature!*

114. Survival International, *Our Land, Our Nature!*

115. Kate Soper, *Post-Growth Living: For an Alternative Hedonism* (New York: Verso Books, 2020).

116. Ashish Kothari, "Conservation Needs Fundamental Economic and Political Transformation," in *Decolonize Conservation: Global Voices for Indigenous Self-Determination, Land, and a World in Common*, ed. Ashley Dawson and Fiore Longo (Philadelphia: Common Notions, 2023).

117. Survival International, *Our Land, Our Nature!*

118. Olúfẹ́mi O. Táíwò, *Reconsidering Reparations* (New York: Oxford University Press, 2022).

119. David Treuer, "Return the National Parks to the Tribes," *The Atlantic*, May 2021, https://www.theatlantic.com/magazine/archive/2021/05/return-the-national-parks-to-the-tribes/618395/.

120. Treuer, "Return the National Parks to the Tribes."

121. Asad Rehman, "The 'Green New Deal' Supported by Ocasio-Cortez and Corbyn Is Just a New Form of Colonialism," *The Independent*, May 4, 2019.

122. Pranab Doley, "The Fight against Colonial Conservation Is a Fight for Millions of People across the World," in *Decolonize Conservation*.

## Conclusion: Climate Debt and Border Abolitions

1.   Brian Osgood, "'Slavery Wages' Prompt Hunger Strike at ICE Detention Facility," Al Jazeera, March 3, 2023, https://www.aljazeera.com/news/2023/3/3/hunger-strike-at-ice-detention-facilities-protest-slavery.

2.  Osgood, "'Slavery Wages' Prompt Hunger Strike."

3.  Detention Watch Network, "Mass Hunger Strike at ICE Detention Facility in Louisiana – Detained People Demand to Be Released," March 2, 2023, https://www.detentionwatchnetwork.org/pressroom/releases/2023/mass-hunger-strike-ice-detention-facility-louisiana-detained-people-demand.

4.  Detention Watch Network, "Immigration Detention 101," https://www.detentionwatchnetwork.org/issues/detention-101.

5.  See, for example, Tom Dreisbach, "Exclusive: Video Shows Controversial Use of Force inside an ICE Detention Center," NPR, February 6, 2020, https://www.npr.org/2020/02/06/802939294/exclusive-video-shows-controversial-use-of-force-inside-an-ice-detention-center; and Freedom for Immigrants, "Widespread Sexual Assault," 2017, https://www.freedomforimmigrants.org/sexual-assault.

6.  Spencer Woodman et al., "Thousands of Immigrants Suffer in US Solitary Confinement," International Consortium of Investigative Journalists, May 21, 2019, https://www.icij.org/investigations/solitary-voices/thousands-of-immigrants-suffer-in-us-solitary-confinement/.

7.  Center for Victims of Torture, *Arbitrary and Cruel: How U.S. Immigration Detention Violates the Convention against Torture and Other International Obligations*, 2021, https://www.cvt.org/sites/default/files/attachments/u93/downloads/arbitrary_and_cruel_d5_final.pdf.

8.  Freedom for Immigrants, *Trafficked and Tortured: Mapping Immigrant Transfers*, February 2023, https://www.freedomforimmigrants.org/trafficked-and-tortured-report.

9.  Freedom for Immigrants, *Trafficked and Tortured*, 27.

10. Freedom for Immigrants, *Trafficked and Tortured*, 3.

11. Elena Hodges, *Building Power: Charting Recent Victories in the Movement to End Immigration Detention in the United States*, June 2022, https://nydignitynotdetention.org/wp-content/uploads/2022/06/Building-Power_Charting-Recent-Victories-Report.pdf.

12. Daniel Zawodny, "Maryland Lawmakers Passed Dignity Not Detention to Protect Immigrants. So ICE Detains Them Elsewhere," *Baltimore Brew*, July 28, 2022, https://www.baltimorebrew.com/2022/07/28/maryland-lawmakers-passed-dignity-not-detention-to-protect-immigrants-so-ice-detains-them-elsewhere.

13. Jane McAdam, *Climate Change, Forced Migration, and International Law* (New York: Oxford University Press, 2014).

14. I follow De Genova, Garelli, and Tazzioli in mobilizing the term *refugee* as a strategic essentialism that refuses the bureaucratic distinction

between undeserving migrants and "genuine" refugees. See Nicholas De Genova, Glenda Garelli, and Martina Tazzioli, "Autonomy of Asylum? The Autonomy of Migration Undoing the Refugee Crisis Script," *South Atlantic Quarterly* 117, no. 2 (April 2018): 247.

15. Stuart Hall et al., *Policing the Crisis: Mugging, the State, and Law and Order* (London: Macmillan, 1978), vii.

16. Michael Hale Williams, *The Impact of Radical Right-Wing Parties in West European Democracies* (Basingstoke, UK: Palgrave Macmillan, 2006), 60–61.

17. See Sunaina Maira, "Radical Deportation," in Nicholas De Genova and Nathalie Peutz, eds., *The Deportation Regime: Sovereignty, Space, and the Freedom of Movement* (Durham, NC: Duke University Press, 2010), 295–328.

18. De Genova, Garelli, and Tazzioli, "Autonomy of Asylum?" 253.

19. Maira, "Radical Deportation," 302.

20. Andreas Malm and the Zetkin Collective, *White Skin, Black Fuel: On the Danger of Fossil Fascism* (New York: Verso, 2021), 133–80.

21. Quoted in Malm and the Zetkin Collective, *White Skin, Black Fuel*, 151.

22. Garrett Hardin, "The Tragedy of the Commons," *Science* 162, no. 3859 (December 1968): 1243–48.

23. UNHCR, "A Record 100 Million Forcibly Displaced Worldwide," May 23, 2022, https://news.un.org/en/story/2022/05/1118772.

24. Internal Displacement Monitoring Centre, "Displacement Data," www.internal-displacement.org/database/displacement-data.

25. Internal Displacement Monitoring Centre, "Displacement Data."

26. UNHCR, "Figures at a Glance," www.unhcr.org/about-unhcr/who-we-are/figures-glance.

27. CARE, "Developed Nations Hugely Exaggerate Climate Adaptation Finance for Global South," January 21, 2021, https://www.care-international.org/news/developed-nations-hugely-exaggerate-climate-adaptation-finance-global-south.

28. Chi Xu et al., "Future of the Human Climate Niche," *Proceedings of the National Academy of Sciences* 117, no. 21 (2020), www.pnas.org/cgi/doi/10.1073/pnas.1910114117.

29. Quoted in Jonathan Watts, "One Billion People Will Live in Insufferable Heat within 50 Years – Study," *The Guardian*, May 5, 2020.

30. World People's Conference on Climate Change and the Rights of Mother Earth, "Final Conclusions Working Group 6: Climate Change and Migrations," April 30, 2010, https://pwccc.wordpress.com/2010/04/30/final-conclusiones-working-group-6-climate-change-and-migrations/#more-1769.

31. Todd Miller, with Nick Buxton and Mark Akkerman, *Global Climate Wall: How the World's Wealthiest Nations Prioritise Borders over Climate Action* (Amsterdam, The Netherlands: Transnational Institute, 2021).

32. See, for instance, Nicholas De Genova, *The Borders of "Europe": Autonomy of Migration, Tactics of Bordering* (Durham, NC: Duke University Press, 2017); and A. Naomi Paik, *Bans, Walls, Raids, Sanctuary: Understanding U.S. Immigration for the 21st Century* (Berkeley: University of California Press, 2020).

33. For a more extensive exposition of these three ecological contradictions, see my book *Extinction: A Radical History* (New York: OR Books, 2016); and Joel Koven, *The Enemy of Nature: The End of Capitalism or the End of the World* (London: Zed Books, 2007).

34. David Harvey, "Why Marx's Capital Still Matters," *Jacobin*, July 12, 2018.

35. Michael Gerrard, "America Is the Worst Polluter in the History of the World. We Ought to Let Climate Change Refugees Resettle Here," *Common Dreams*, June 26, 2015, https://www.commondreams.org/views/2015/06/26/america-worst-polluter-history-world-we-should-let-climate-change-refugees-resettle.

36. For a fully fleshed out defense of border abolition, see Gracie Mae Bradley and Luke de Noronha, *Against Borders: The Case for Abolition* (New York: Verso, 2022).

37. George Monbiot, *Out of the Wreckage: A New Politics for an Age of Crisis* (New York: Verso, 2017).

38. Mark Fisher, *Capitalist Realism: Is There No Alternative?* (Winchester, UK: Zero Books, 2009).

39. Cormac Cullinan, "The Universal Declaration of the Rights of Mother Earth: An Overview," in *The Rights of Nature: The Case for a Universal Declaration of the Rights of Mother Earth* (San Francisco: Global Exchange, 2011), 10.

# INDEX

# ABOUT THE AUTHOR

Ashley Dawson is Professor of English at the Graduate Center / City University of New York and the College of Staten Island. He is the author of several books on key topics in the environmental humanities, including *People's Power: Reclaiming the Energy Commons*, *Extreme Cities: The Peril and Promise of Urban Life in the Age of Climate Change*, and *Extinction: A Radical History*. A member of the Public Power NY campaign and the founder of the CUNY Climate Action Lab, he is a longtime climate justice activist.

Printed in the USA
CPSIA information can be obtained
at www.ICGtesting.com
JSHW012016230424
61720JS00004B/96

9 798888 900581